PICTORIAL GUIDE TO MAMMOTH CAVE, KENTUCKY

By

Reverend Horace Martin

Illustrated in the First Style of Art By:

S. Wallen, Jno. Andrew,

J. W. Orr, and N. Orr.

PHILADELPHIA:

KING & BAIRD, PRINTERS, No. 9 SANSOM STREET.

Table of Contents

AUTHOR'S ADDRESS

In America, Nature seems to have purposely operated on a gigantic scale. Her Lakes, her Rivers, and her Mountains, may be instanced as an attestation of what we say. Greatness and sublimity characterize them all. Poets have sung their praises, tourists have described them in all the eloquence of prose, and painters have labored to illustrate them upon the canvas. They have been famed everywhere. But at the same time, an object of Nature, as sublime as beautiful, and as great as the Andes or the Mississippi, has been comparatively neglected, we allude to the MAMMOTH CAVE of Kentucky. Little has hitherto been said of it by authors, less done towards familiarizing it to the Million, by the painter. It is with the view of filling up this blank, though it may be imperfectly, that the present work has been undertaken.

We are anxious that Americans, as well as foreigners visiting the Republic, should view the Mammoth Cave; and we are aware of no better plan of insuring this, than by describing the several features of attraction, as well as incidental comforts, which a trip there is sure to include. With the facilities in the way of travel, which exist in this country, no excuse of distance can be offered. A thousand or more miles can be covered, with no more harass or fatigue than that consequent upon quitting one train for another, or the railroad car for the steamboat. And what scenery will the eye of the traveller command in his transit? and what spectacles of splendor await him at his journey's end, where the Mammoth Cave itself is situated?

There we will leave our friends, but perhaps not entirely so, as we shall be with them in the representative form of this, their Guide and Companion. That the association may not be deemed an unworthy one, we most sincerely hope; while the faithfulness of our delineations, (so far as the Cave is concerned,) and of our testimony to the elegant care and "home comfort" to be found at the neighboring hotel, cannot, we feel, be impugned by any one.

Entrance to Mammoth Cave

Cave House

MADE. JENNY LIND'S VISIT TO THE MAMMOTH CAVE

SAVORED BY JULIUS BENEDICT, ESQ.

Early in the morning, after the close of our last concert in Nashville, we started, with a somewhat smaller party than had hitherto accompanied us, on a trip to the Mammoth Cave. It consisted of Mademoiselle Lind and Mademoiselle Ahmansen, Belletti, M. Hjortzberg, Mr. Burke, Mr. Seyton, and myself. Our road was rough, and in many instances almost impassable for a carriage. The rain had however, laid the dust, and although there was little of the picturesque to be met with in the country that was stretched on either side of us, the fresh brilliancy of the young year sheeted trees and meadows alike in its budding green. After partaking of luncheon at Teyress' Springs, and pausing to dine afterwards at the Bowling Green, we arrived in the course of the evening, and found ourselves in comfortable lodgings, at Bell's Hotel. On the following morning we quitted this tarrying place at nine o'clock, and had the satisfaction of travelling over eight miles of the very worst road we had yet traversed in the United States, charmingly broken up with snatches of woodland and forest scenery here bending past the edge of a jagged and abrupt glen, and then breaking into a sweep of meadow or budding foliage. At length we arrived at the hotel, a dismal and queer looking building, the roof of which was seamed with the chance skylights made by age and decay, and the service of which was performed by domestics, who were scrupulously bent on following their own fancies in the management of our table, for here it was that we breakfasted. In truth, the meal itself was excellent, and the room in which it was held, considering the time of the year, was in good order, Jenny Lind's presence, we presume, having, as is usual in hotels, railways, and steamboats, made an extra season.

Fortunately, we here met with Mr. Croghan, the proprietor of the estate in which the Cave is situated, a most gentlemanly and delightful person, who did us the honors of his subterranean dominions in the most agreeable manner. It was about twelve o'clock that we started in his company for the Cave, and to avoid the pertinacious curiosity of the guests, who had been collected here by the report of Mademoiselle Lind's visit, he conducted us by a less frequented pathway than the one usually taken to its mouth. Lamps were now procured, and as it happened, we were fortunate enough to be placed in the hands of the very Prince of Guides. This was Stephen, who must be a well known character to those who visit this palace of the Gnomes. Half Indian and half negro, (a singularly rare mixture of blood,) he has been living in or about this cavern for the last fifteen years, until he himself has begun to fancy it would be impossible to quit it. Although of course, uneducated, he is essentially a clever man, and has contrived to pick up a vast amount of information from associating with every description of persons. Now he sports a bit of science, derived from some of the more learned visitors he has conducted through the cavern, or a bit of artistic knowledge which has been dropped behind him by some wandering painter, or haply a touch of the life of the world beyond, which has filtered through his mind from a thousand sources. In addition to this, when it is remembered that he is as much at home in the lengthy avenues, the gorgeous churches, and palatial halls, the domes and the pits of this weird region, as if he had been amongst them, it must be admitted that it would be somewhat difficult to find a guide better calculated to do its honors. To give anything approaching a thorough description of the cavern, is far from our purpose. Indeed, we shall be well satisfied if we can but impress the reader with the conception that masters our own sense, as we take the pen in our hand with the vain hope (for we cannot but feel that it is so,) of doing something like justice to the effect it produced upon our minds. In fact, it is and would be well nigh impossible to give with pen and ink any idea of the wondrous effect and extraordinary combinations of nature's architecture, with her wondrous and delicate tracery which strike the visitor at every step he takes in these intricate and winding labyrinths. Now you enter what I would appear to be the sacred precincts of a Gothic Chapel. What is visible of the roof as the light of the flashing torches is caught upon it is seamed with arches. Elaborate pillars wreathed with tracery cluster along its sides. The very pulpit is chased with elaborate and tangled ornaments, and appears ready for the preacher. After this, you bend your way through a rough and tedious path, that winds through fragments of rock, and fallen masses of rough and jagged stone. This brings you to a wooden bridge, over which you pass, and reaching apparently the side of the cave, gaze through a broken space into the thick and heavy darkness beyond it. Here the glaring lustre of a Bengal light, touched by the torch of Stephen, falls into and for some moments partially illumines the profound depths of a place which is called the Bottomless Pit; and, indeed, nothing could well give a more vivid idea of the earthly entrance to a spiritual Hades than does this place. The spot of intense and glowing light, the unfathomable space below, the unnatural features of the place, all brought out in strong relief by the unusual radiance; and the awful silence that reigns around, unbroken, save by the whispers and muttered observations of the party which stands almost lost in the gloom of the silent cavern, give it a character of extreme and unutterable solemnity. What, however, must we say of the 'Star Chamber'? After having wandered for a mile or more along what we presumed was the principal avenue, (the height of this varies, as we should suppose, from thirty to eighty feet,) we passed the 'Giant's Coffin,' a mass of stone presupposed by the dealer in fabulous nomenclature to be the tomb of some antediluvian hero. Here the Cave widened, and we found ourselves standing as we seemed to emerge from it, under a broad and sable sky, spotted with unknown stars. Almost for the first moment you might dream that you had entered upon another world. The illusion is complete. Above you lies the vault of the dark and novel heaven, seamed with apparently countless planets, and around you stretches the dark and weird looking horizon, apparently dying away into the gloom of that strange firmament. Here also our guide shone in all his glory. First he would withdraw within the entrance, carrying the torches with him. Then the stars would disappear, one by one, until we were left in silence and darkness. Anon a crimson light would break out among the rocks, whose intense brilliancy would give us some idea of the grandeur and splendid proportions of the 'Star Chamber,' sparkling in its brilliant glory on the glistening spots of the sable coping. Then he would descend and move further off; to throw the light of the torches on others of the incrustations and glistening stalactites of the Hall. Suddenly the notes of a violin were heard breaking on the stillness, and the Prayer from the *Der Freyshutz* poured its melody on the Chamber. For a moment we were so struck by the unexpected sounds, that we barely looked at each other. Soon we, however, began to notice that Burke was

9

absent, and remembered that he had brought his instrument with him on this trip. The mystery, a rare and delicate one, too, was unravelled. After leaving this spot, we passed through the 'Fat Man's Misery,' and the 'Happy Relief,', which last, we confess, we should have presupposed to have been achieved only by a course of sudorifics, and at last reached the borders of Lethe. Unluckily there was no Charon in waiting to bear us across the ominously named stream. This may possibly appear an anomaly, yet when it is known that the grandest part of the Cave lies beyond Lethe and Styx, our mortification at finding the first river impassable, and the Tartarus beyond it, out of our reach, may very readily be conceived. The waters had unfortunately risen so high during the last few weeks, that the impossibility of passing the streams of this subterranean Tophet was self evident. We were therefore compelled unwillingly to satisfy ourselves with the glowing reports of Stephen and Mr. Croghan, of the 'Crystal Chambers,' the 'Echo Halls,' stalactitic Domes, fishes without eyes, and rats that were half rabbits, with sundry other breathing and visual oddities that were to be found in these infernal regions. We were, however, richly repaid for our visit by that which we had already seen, and the crowning point was our pause on our return beneath Goran Dome. Fancy an immense wall in the bowels of the earth, lit up as if by magic, (i. e. Bengal lights,) with its carved cornices and sculptured or arabesqued architraves, coming crisply and exquisitely off in the momentary and brilliant lustre, the wall standing some 400 or 500 feet in height, and the silence of the scene being only broken by the slow dripping of the water which trickles through the interstices of the rock. Possibly the only disappointment which was induced in my mind, arose from the width and breadth of the chamber not corresponding with the height, which had it done, the impression of grandeur given by this singular scene would have been quadrupled! At length, after having roamed about without a moment's rest for more than four hours, in which time we had not explored the twentieth part that is already known of this splendid palace, reared, as it might almost seem, for another race of beings, we passed through the 'Bats Chamber,' where thousands of these creatures remain, as though they were spellbound, hanging to the walls in their winter sleep, and, emerged again into the world above us, which seemed to fasten upon our senses with an almost crushing weight, as we found its light dimming and blinding the eyes which instinctively sought the radiance of the sunny heaven. We returned to the hotel, dined there, and bidding a kindly farewell to Mr. Croghan, were soon on our way to the place at which we were that night to sleep, carrying with us a recollection which will not readily be effaced either from the mind of Mademoiselle Lind or of myself.

GUIDE TO THE MAMMOTH CAVE

Among the many grand and beautiful objects of Nature which have been worthily described, either by the pen or the pencil, or by both, the Mammoth Cave in Kentucky cannot be numbered, the as yet published details or pictures of it failing in imparting anything like an accurate idea of its many wonders. Having recently visited the Cave, and having all our feelings of admiration warm within us, it seems that the few tourists who have hitherto given to the world their ideas of this splendid natural object, have, by constitution, lacked that sense of the beautiful which is the readiest interpretor of scenes like those which it is our purpose to describe: hence, we have had nothing more than a few dry details, where the true magnificence of the theme is of a character to inspire the pen to something much higher than commonplace.

By many who, however, are gifted with the quality, the lack of which in others we have just noticed, the Mammoth Cave has been designated the "Eighth Wonder of the World." Some, indeed, have gone so far as to claim for it precedence before the "seven": and it would be difficult to decide whether it has not claims to such a proud preeminence. So widespread is its fame, principally from the verbal reports of those who have visited it, that thousands of American citizens, as well as strangers from every other part of the world, make a point of going to it. We have had opportunities of witnessing its effect upon persons of different temperaments. In the religious it has awakened devotional feelings amounting almost to delirium; in the poetic or artistic impromptu bursts of eloquence and a sharpened sense of the grand; the singer, too, has felt the exalting and purifying influences of the place, and has hymned his or her praises to the power there so obviously manifested. Few have shown insensibility; and they, we hold, to be objects of pity.

Persons in the eastern states, desirous of visiting the Mammoth Cave, had best take the steamboat from Cincinnati to Louisville: the distance between the two places is 133 miles, and the fare only $2.50. The boats of this line are first class ones, and possess every recommendation; their arrangements are so complete, that were it not for the noise of the engines, a person might fancy himself in a fashionable hotel. The distance is performed in a very short period of time. Arriving at Louisville, the tourist should take the stage, which will convey him ninety miles over a turnpike road, and stop at a truly comfortable inn, kept by Mr. Bell for the last twenty five years. The stage is now left to pursue its way to Nashville. Parties coming west would take the stage from Nashville, and stop at Bell's. There are, besides, steamers which ply between Louisville, up the Green River, to within one mile of the Cave. In these, our preliminary details, it may be as well for us to state that a company is now forming, whose object is to make a railway from Nashville to Louisville, which, when completed, will pass within ten miles of the Cave. Proper conveyances will be in readiness at that distance, to take persons directly to the Cave itself. The stages now running on the road take the mails, and may be relied on for their punctuality, always arriving at Bell's about half past 9 o'clock at night. Every accommodation for the night will be afforded, and in the morning, when the visitors are quite ready, they are sent over to the Cave, in either a carriage or a coach, according to their numbers. The distance from Bell's to the Cave is nine miles, and the road runs through a very romantic country, now ascending lofty mountains, then traversing thick woods, which the city visitor will find strongly to contrast with the more familiar scenes of his life. It is supposed that the Cave runs under the ground traversed between its mouth and Bell's, as high rocks are frequently met with. Descending from the last hill, we arrive at the entrance to the grounds, and in sight of the hotel. This property is of vast extent, containing more than 1700 acres of ground; the proprietor has purchased all the land, under which it has been ascertained the Cave traversed its many winding ways. The grounds, which the visitor now enters, are laid out with consummate taste displaying ornamental trees and much shrubbery, with forest trees of great antiquity. Having traversed a winding avenue, the tourist at length arrives at the Cave Hotel, conducted by Mr. Miller, who is a very respectable, accommodating, and affable gentleman, and ever ready to please his company to the utmost extent. The hotel consists of a number of buildings of different dates, having been increased from time to time, to meet the continually increasing number of visitors. The establishment, in its present state, is capable of lodging 150 visitors, and during the season always has that number. The terms are very moderate; and with a travelling experience which extends over the whole United States, we can safely say that we are acquainted with no place at which a person who wishes to withdraw himself from busy life, can find greater comfort, or can enjoy Nature in her unadorned loveliness, more than here. There is no town or village within twenty miles, and neither the one nor the other contains more than a comparatively thin population. Consequently, visitors are not annoyed by the influx of the rabble on certain days of the week. In this beautiful and retired spot, the stranger will meet with polished and refined society, persons from all parts of the world meeting there. At the hotel a register is kept. Some celebrated names are inscribed on its pages, among them, those of Jenny Lind, Jules Benedict, Belletti, etc. The former gentleman has favored us with a sketch of his and companion's visit to the Mammoth Cave, which we append to this work. The hotel is two stories high, and two hundred feet long, with brick buildings at each extremity, showing their gable ends in front. The space between is occupied by a long wooden building, with a piazza, and gallery over it. At the end of the hotel runs a long row of log houses, one story high, with colonades in front, the whole length, which must be near two hundred feet. These colonades and piazzas must be very convenient in wet weather, helping to form as they do, a beautiful promenade, protected from the rain or sun. The dining room of the hotel is a spacious apartment, while the fare displayed upon its table of the finest quality. Venison is always to be found here in abundance. A large kitchen garden is kept in a high state of cultivation, to furnish vegetables and fruits for the table; several gardeners are retained in the establishment. Over the dining room is an apartment of precisely the same dimensions, fitted up for a ball room; and an excellent band is kept during the entire season, for the purpose of amusing the visitors. In another part of the premises a ten pin alley is fitted up. Indeed, taking the whole arrangements of the hotel, we cannot speak too highly of them. The perfect comfort of the visitor is the proprietor's evident aim. There are single as well as double bedrooms in different parts of the building; the log houses are intended for families, as each little house is fitted up for one, which can live as privately as possible, or mix with the general company, whichever seems the most agreeable. Many of the parties lodged in the hotel are in the habit of exploring the Cave more than once, frequently perhaps, before they can acquire even a partial knowledge of it. In this case, the rule is, that they pay their entrance or cave fee once, and so often as the guides go in with fresh visitors, the old visitors have the privilege of accompanying them, without being required to pay any second fee, so that a particular party may visit the Cave a hundred times, and yet only pay one fee. Persons form themselves into companies, each day, to hunt, or fish, as well as to visit the Cave.

And now for the Cave, well and aptly designated Mammoth; and as a natural object, perhaps unequalled by any other in the world, certainly unsurpassed.

In order to explore only one of its avenues, which is nine miles in extent, the visitor starts immediately after breakfast. The entrance to the Cave is about two hundred yards from the back of the hotel. Leaving it, the expectant tourists pass down a beautiful ravine, having on each of its sides towering trees, their foliage forming a beautiful arch overhead, so umbrageous as to shut out all vision of the blue sky. About the trees grapevines are entwined, and flourish in luxuriance. sharply round to the right hand, the visitor approaches the entrance of the Mammoth Cave.

Entrance to the cave, viewed from inside

For a painter, the scene now presented would make a splendid study. It is difficult, in fact, to find words sufficiently expressive to describe the beauty of this spot. Descending gradually to the bottom of the dell, and turning

He is now under its arch, having made a descent of some thirty feet of rude stone steps. Before him is a small stream of water. It falls from the front of the crowning rock, its sound being wild and unequal. The ruins below receive it, and it ultimately disappears in a deep pit. Let the visitor now look backwards. How awful must be his sensations! All is utter gloom; and well may he exclaim "This is chaos!"

The visitors are invariably accompanied by a guide, who places in each of their hands a lamp, which he furnishes with oil from a canteen swung round his back. The journey underground is then commenced.

Hoppers and Water Pipes.

The first objects that now attract attention are the wooden pipes which conducted the water, in its descent, to the hoppers, when the Cave was used for the manufacture of saltpetre. In speaking of this part of the Cave, we are reminded of an incident of which it was once the scene.

We refer to the discovery of a skeleton, so peculiar in its proportions as to suggest the existence, at some remote period, of that giant race of men, of whom there has been so much said in works generally believed fabulous. This skeleton was found some years since by the saltpetre workers, and was buried by their employer in the spot where it now lies. This was done to quiet the superstitious fears of the men. Some antiquary had, however, taken the precaution to possess himself of the head. Another relic of the olden time was found in the Cave. It was a bow, precisely similar to those worn by the Red Indians, and no doubt, had belonged to one of that race when they were the sole masters of the soil, many ages, indeed, before the white men had come among them, to drive the aborigines to the deep forests, and to alter the destiny of this mighty continent by the agencies of civilization.

Let us return to the visitors. Having proceeded onward, a doorway is reached. It is set in a rough stone wall, stretched cross ways, and so blocking up the entire cave. We proceed through this passage, which is called "the Narrows" for a short distance;

then making a gradual descent, our friends find themselves in the great vestibule or antechamber of the cave. How awful is now the surrounding darkness! No where can there be discovered the least glimmering of light. The eye searches for it, but in vain. Blackness reigns. It is under, above, around you, There is one way, however, to dispel it. You are told the way, but for a few minutes you are credulous. Presently you make trial, and then how wonderful is the change! More than a hundred feet above your head is a gray ceiling, moving away, so it seems, majestically and spectral. Then your sight is cognizant of buttresses. They seem tottering, and as though superinduced to the action by the immensity of their upward weight. All continues silent, and the sensation in the brain is that of a tingling agony. The guide, who knows the power of antithesis, and is an adept too, in dramatic effect, now lights a few fires, by the aid of which you ascertain that you are in a basilica of an oval form, of some two hundred feet in length by one hundred and fifty in width, with a flat and level roof between sixty and seventy feet high. At right angles are two passages, opening into this huge chamber. They are at its opposite extremities, and in consequence of their preserving a straight course for five or six hundred feet, with the same flat roof just particularized, the impression is, that they are included in one magnificent hall. The passage on the right hand is designated the great Bat Room. That in the front, is the beginning

21

of the Grand Gallery, or the Cave itself. The reader will be surprised when he is told that the entire space is covered by a single rock, in which it is impossible to discover a break or an interruption, if we except the borders, where is a broad sweeping cornice, traced in hori zontal paved work. And what will also appear extraordinary, not a single pillar contributes to the support.

Vast heaps of nitrous earth, and of, the fragments of the hoppers, which latter are of heavy planking, encumber the floor. Before the Cave was appropriated to the manufacture of saltpetre, it was used as a burying ground; and it is said that the workman disinterred many a skeleton belonging, it is presumed, to that gigantic race of which the bones mentioned in a previous page were a specimen.

Those visitors who enter the great Bat Room, will have some suspicion that they are passing into infinite space. This impression will continue some time. The walls of the cave are so dark as not to admit a single reflection from the torches carried on ordinary occasions. It is possible that a greater number would have power to dispel the gloom, and to enable the explorers to form some idea of its grandeur.

The Audubon Avenue (in which we must now fancy the visitor to be) is more than a mile long, fifty or sixty feet wide, and as many high. The appearance of the roof is mystical or grand, exhibiting, as many other parts of the Cave do, a kind of floating cloud. Of late years, a natural well, twenty five feet deep, containing the purest water, has been discovered. Round it are innumerable stalagmite columns. Springing from the floor, they extend to the roof, and bear incrustations which, when lights are held or suspended near them, reflect a thousand rare and beautiful objects.

The Little Bat Room Cave may be designated a branch of Audubon Avenue. It is on the left as the visitor advances, and in distance, is not more than three hundred yards from the Great Vestibule. In length it is little more than a quarter of a mile. It is remarkable for its pit, which is two hundred and eighty feet in depth; and is also noticeable as being the hibernal resort of bats. Tens of thousands of those ominous looking birds are seen hanging from the walls, seemingly dead or torpid during the winter, but when spring comes, the place knows them no more.

The visitor will now pass for a second time through the Vestibule, and enter the main Cave or Grand Gallery. This is a mighty tunnel, and extends for many miles. Its average dimensions throughout, are fifty feet in width by as many in height. It is a truly magnificent avenue, crowded with objects of interest, and the largest, we believe, in the world.

About a quarter of a mile down the main Cave, the visitors arrive at what are called the Kentucky Cliffs. The name, it is said, is derived from their imagined resemblance to the cliffs on the Kentucky River. Further on, at a descent of some twenty feet, is the Church. This part of the Cave is sometimes reached by a gallery from the cliffs. The way we have indicated is, however, the one generally adopted. The ceiling of the Church is sixty three feet high, and the Church itself, taking in the recess, is not less than one hundred feet in diameter. It is furnished with a natural pulpit, and as though to perfect its resemblance to the place whose name it bears, there is a hollow behind the pulpit, in which an organ might be very well placed, not to say anything of a full choir. Thousands can be very well accommodated in this Church, and, indeed, sermons have been preached there, before very large congregations. In its capacity for sound, it is as well fitted for a place of public worship, as in its other characteristics. Even a very slight effort on the part of a speaker would render him thoroughly audible to those seated furthest from him. Standing in this natural temple of God, older by far than any cathedral the eye ever looked on, and as beautiful, the imagination conjures up the scene it would present were it again to be appropriated to service, and the proper ceremonial be observed. The surpliced priests, the listening auditory, the organ's swell, the burst of voices, the form of the Church itself, and the "dim, religious light," all that tends to the completion of an ideal, fascinating in the highest degree, and not easily dispelled in a place like that of which we speak.

Concerts as well as religious services have been held here; and we have no doubt but that the very inspiration of the place has awakened melody until then unknown and unappreciated.

Quitting the Church (which the visitors will not be very anxious to do) a large embankment of lixiviated earth will be seen. It was thrown out by the miners more than thirty years ago. The former uses of the place are further indicated by the print of wagon wheels and the tracks of oxen, both as sharply defined as though they had been made but a few weeks since. Proceeding a little further, the visitor arrives at what is called the Second Hoppers. Here, too, are the signs and tokens of other days, and of uses different to those which are identified with the place at present. Here are the ruins of old nitre works, leaching vats, pump frames, and two lines of wooden pipes; one to lead fresh water from the dripping spring to the vats filled with the nitrous earth, and the other to convey the lye drawn from the large reservoir, back to the furnace at the mouth of the Cave. It has been stated, on authority, that the nitrous earth in the Mammoth Cave is, in itself, sufficient to supply the whole world with saltpetre. Previous to rejoining our friends in their explorations, we may perhaps be pardoned for giving the following particulars, quoted from an official document:

"The dirt in this Cave gives from three to five pounds of nitrate of lime to the bushel, requiring a large proportion of fixed alkali to produce the proper crystallization, and when left in the Cave, becomes reimpregnated in three years. When saltpetre bore a high price, immense quantities were manufactured at the Mammoth Cave; but the return of peace brought the saltpetre from the East Indies in competition with that of America, and drove the latter entirely from the market. An idea may be formed of the extent of the manufacture of saltpetre at this Cave, from the fact that the contract for the supply of the fixed alkali alone for the Cave, for the year 1814, was twenty thousand dollars.

"The price of the article was so high, and the profits of the manufacturer so great, as to set half the world gadding after nitre caves, the gold mines of the day. Cave hunting, in fact, became a kind of mania, beginning with speculators, and ending with hairbrained young men, who dared for the love of adventure the risk which others run for profit. 'Every hole,' remarked an old miner, 'the size of a man's body, has been penetrated for miles round the Mammoth Cave, but although we found petre earth, we never could find a cave worth having.' "

The visitors are now looking from the nitre works towards the left, and perceive, thirty feet above, a large cave. In connexion with it is a narrow gallery sweeping across the main Cave, and losing itself in another cave, which is to the right; the latter is called the Gothic Avenue. And there is every evidence to show that it was at one time connected with the cave opposite, and on the same level, forming, as it were, a bridge over the main Avenue, but subsequently broken down by some natural convulsion. The Gothic galleries are extremely beautiful.

The Cave on the left is filled with sand, and has been penetrated a short distance only. Most likely, however, from its ample size at the entrance, it would, were all obstructions taken away, be found to have an extent of some miles.

The visitor on leaving the main Cave, and ascending a flight of steps of about thirty feet, will find himself in the Gothic Avenue. This portion of the Cave is so named from its strong resemblance to a Gothic building. Its dimensions are, in width forty feet, height, fifteen feet, length two miles. Nothing can be more smooth than the appearance of the ceiling; in fact, it seems as though the artizan had given it the last touch, and it was only waiting the process of drying. An excellent road has been made in this Cave; and the atmosphere is temperate.

Entrance to the Gothic Gate

At an elevation of a few feet above the floor, and fifty feet from the head of the stairs, leading up from the main avenue, a couple of mummies were to be seen in the early part of the present century. At the time we speak of they were in excellent preservation. One was of a female, and various articles of the wardrobe were by the side of it. Of the ultimate fate of these mummies nothing can be said with any approach towards certainty. One, we have been told, was destroyed in the burning of the Cincinnati Museum; while the wardrobe of the female, we have also been given to understand, was given to a Mr. Ward, of Massachusetts, and afterwards presented by him to the British Museum.

A third mummy was found by the miners, in Audubon Avenue, in the year 1814. It was their wish to conceal it; accordingly, they placed large stones over it, and fixed such marks about it as to direct them to it at some future day. But all the pains they took were in vain. In 1840 the present keeper of the hotel, Mr. Miller, having ascertained the facts as we have related them, set out in search of the place described to him as the grave of the mummy. He soon found it out by the aid of the lights he had brought with him; the stones, however, which had been placed above it, had so injured it, as to leave it of little value to the antiquary or any other person. It is no improbable hypothesis, we think, that were judicious efforts made, several mummies would be yet found, and also other objects which would tend, in no slight degree, to throw light on the early history, not of the Cave alone, but of the great continent itself, of which it forms so conspicuous an object of interest.

The allusion to mummies very naturally reminds us of the discovery of one in this cave, some years since, and a few particulars of which may not be uninteresting to the reader.

A scientific gentleman, one of the earliest visitors of the Cave, saw the above named relic in the year 1813. While digging saltpetre earth in the short cave, a rock somewhat flat was met with. It was a little below the surface of the earth. The stone, which was about four feet wide by as many long, was raised, and discovered a square excavation about three feet in width, the same in length and depth. On examination, it proved to be the sepulchre of a human being, a female, with her wardrobe and ornaments placed near her. The body, which was found in a sitting position, was in a perfect state of preservation. The arms were folded up, and the hands laid across the bosom; while around the two wrists was wound a small cord, apparently for the purpose of keeping them in the position originally determined. Round and near to the body, were wrapped two deer skins, which, on examination, seemed to have undergone a mode of dressing very different to any known at that day, and which would have been equally inexplicable now. The hair had been cut off very near the surface of the skins, which were ornamented with imprints of vases, leaves, etc., sketched with a substance quite white. On the outside of these skins was a large square sheet. It was neither wove nor knit, the fabric being of the inner bark of a tree, which, it was supposed, from certain appearances, was that of the linn tree, resembling as it did, in texture, the well known South sea cloth, or matting. The head and body of the mummy were entirely enveloped in this sheet. The hair had been cut off very short, being only the eighth of an inch from the scalp; on the neck, however, it had been allowed to grow one inch. In color it was dark red. The teeth of this mummy were perfectly white and sound. No blemish whatever was discoverable upon the body, with the exception of a wound between two ribs near the back bone; one of the eyes had also been injured. The nails on the toes and hands were long and quite perfect. The features of the face were remarkably regular; and it appeared from measurement of the limbs, that the body, in life, must have been something over five feet ten inches in height. At the time of discovery, it weighed only fourteen pounds. It was perfectly dry, but on exposure to the atmosphere, it gained four pounds in weight, by the absorbing of the dampness.

At the side of the body were found a pair of moccasins, a knapsack, and what in the absence of a more expressive term we must call a reticule. Wove or knit bark was the material of which the mocassins were made; and round the top there was a border, to add strength, or, perhaps, to serve as an ornament. The size of the feet, as denoted by these, must have been very small. In form, the moccasins of this subterranean mummy were precisely the same as that of the articles at present worn by the northern Indians. The knapsack was also of wove or knit bark, having a deep and strong border around the top, and in size was the same as the knapsacks worn by soldiers. In workmanship it was exceedingly neat, and the circumstance of its being so, with the preparation of the fabric itself, were sufficiently indicative of a high degree of skill, as characteristic of a period and a race of which we, the modern generation, have generally so derogatory an opinion. The reticule, as well as what has been already enumerated, was made of wove bark, and in shape, was like a horseman's valise, opening its entire length on the top. It was furnished with two rows of hoops, each row on its respective side near the opening. Two cords were fastened to one end of the reticule at the top, which passed through the loop on one side and then on to the other, the whole length. By these means it was laced up and secured. Deep borders, of a fanciful and pretty pattern, run round and strengthened the top of the reticule. The articles proper to it and the knapsack were: a head cap, made of wove or knit bark, borderless, and in shape, similar to a very plain night cap; seven head dresses, made of the quills of very large birds, and united in the same manner as feather fans are made, with this exception, that the pipes of the quills were not drawn to a point, but spread out in a straight line with the top. This was done by perforating the pipe of the quill in two places, then running two cords through the holes, and finally, winding round both quills and cords, fine thread, so as to fasten each quill in the place designed for its reception. Extending some distance beyond the quills on each side, these cords could be tied on placing the feathers on the head. Among the other articles to be enumerated were some beads, several hundred strings. These consisted of a very hard brown seed, smaller than hemp seed. In each of them a small hole had been made, and through it a three corded thread, somewhat like sieve twine, was drawn. Then there were red hoofs of fawns, on a string, by which they had the appearance of a necklace; the claw of an eagle, having a hole through it, by which a cord could be attached, so as to enable a person to wear the claw; the jaw of a bear, apparently intended to be worn, like the former; two rattlesnake skins, one having fourteen rattles upon it; some vegetable colors done up in leaves; a small bunch of deer sinews; several bunches of thread and twine; seven needles, some of horn, some of bone, and remarkably smooth; a hard piece made of deer skin; two whistles made of cane, and each about eight inches in length.

At the period of the mummy's discovery, various conjectures were formed relative to it. Its preservation in the Cave was no theme for wonder, as the nature of its atmosphere is such as to preserve animal flesh for an indefinite period of time. That many centuries had rolled away since that discolored remnant of mortality moved about the earth, and had thought, feelings, and sensations like ourselves, there could not be a doubt, but the age, the condition, or the circumstances attendant upon death remained a mystery. There were no tokens, or, if there were, not sufficiently suggestive of explanations, to mark the precise date of the world's existence, to which that mummy, as a living thing, could be referred. All was vague and shadowy; and those antiquarians who were interested in the discovery, and would have travelled a thousand miles, to come to any definite conclusion, were obliged to content themselves with this: that the body was that of a human being who might have existed but a few ages after the Flood or the birth of the Saviour, a wide interval indeed, but taken at the latest, amply sufficient to establish the distinction of ancient times.

After this somewhat long, though necessary digression, let us return to our friends, the visitors. They are now in the Gothic Avenue, once called the Haunted Chamber, why they will be informed by the guide as they pass along. As the particulars share so equally the elements of the grave and the gay, we may be excused giving them in a hasty way: A miner, new to the Cave, and therefore not very well versed in the methods of averting or escaping from danger, was sent one day, in company of an older workman, to the Salts Room, for the purpose of digging a few sacks of the required article. Seeing that the path was unobstructed, and that the Haunted Chambers were in a continuous line, the young man, who, by the way, was a little vain, and wished to show off his bravery, declined further direction, and went off quite alone. Several hours passed, and as he had not made his appearance since the morning, his fellow workmen became somewhat alarmed. The circumstance was described, and the young man's vain glory remembered. Some of the miners concluded that the fellow, like many other gentlemen with an exaggerated opinion of their own powers, had his career suddenly stopped; others, however, hoped that he had been spared, to repent of his follies, and to become a wiser man. A consultation was held, and it was, at length, determined to go in search of the missing man. Some six were formed into a company. They were negroes, and previous to starting on their errand of mercy were stripped half naked. It may, therefore, be imagined how extraordinary was their appearance. In the meantime, the young miner, after reaching the Salt Room, filling his sack, and succeeding in getting half way back to the place whence he started, thought it possible he might be going wrong.

His valor vanished before the bare idea. It was no longer his courage and his clearness that he thought of, but his sins. They rose up before him, and, it may be, that the gloom of the place where he was, and his fear (which is always a marvelous magnifier), made the errors of his youth seem greater and graver than they really were. In his agitation he run here and there, prayed heartily, and cried lustily. Ultimately, he dropped his lamp, which was immediately extinguished, and fell over a rock. He prayed for help when, in his terror, he thought it could alone be rendered; but hours passed away, and all help, even the slightest, seemed as far from him as ever. Madness seized him. He thought that he had quitted earth, was disembodied, in fact, that he was in the place of torments said to be reserved for sinners. He gazed around him. Merciful powers what are those moving figures. He had never seen anything like them. They were spirits, sent to drag him to his punishment. He hears their yells. Were ever mortal voices like the wild outburst ringing in his ears? Never! Nearer and nearer they come. He is conscious of their hot and hissing breath. Their arms are outstretched to clutch him. He will soon be in their embrace, fast locked. Horrible! They have him, they, not devils, but miners like himself. He knows them, and that he is a saved man. "Hurra! hurra! hurra!" Never did the Mammoth Cave, Old Kentucky, reveberate with such shouts as on the me morable occasion of the miner's rescue.

Exclamations of wonder and delight, too, are indulged in by the visitors to the Cave, when they view the stalag mite columns, reaching from the floor to the ceiling of the avenue in which we left them. They are, however, soon startled by a hollow and ominous noise, the echo of their footsteps; but the guide soon quiets their apprehensions, by the assurance that the noise is to be attributed to the proximity of another large avenue underneath, which, strange as the assertion may appear, has been frequently visited. Near this place are a number of stalactites. One bears the name of the "Bell," and on being struck, it formerly sounded like the deep bell of a cathedral. But of late years it has been quite mute, having been broken by a tourist from Pennsylvania, who, we presume, was partial to loud music.

Louisa's Bower and Vulcan's Furnace are now passed. They are marked by a heap of something similar in appearance to cinders, and some dark colored water. The New and Old Register Rooms are the next in succession. The ceiling of these rooms is perfectly smooth, and would be of an unsullied white, had not numerous persons, with a disgusting egotism, traced their obscure names upon it with the smoke of a candle.

After the visitor's departure from the Old Register Room, it is the usual practice of the guide to receive all the lamps, with the exception of one, which of course, is necessary for the purposes of exploration. A place of surpassing magnificence is then entered. Here, as in many other portions of the Cave, language will be found inadequate to describe what is to be seen. The hall (if we may call it so) is elliptical in form, in dimensions, eighty feet long by fifty feet wide. The two ends are nearly blocked up by stalagmite columns of large size. Two rows of pillars, smaller than the others, reach from the floor to the ceiling. They are equidistant from the wall on either side, and extend the full length. Now the purpose of the guide in requesting the visitors to give up their lights is apparent. He has, in the interval of departure from the Old Register Hall to the arrival in the present portion of the Cave, so disposed the lamps as to cause their reflections to fall upon the pillars and ceiling, indeed, upon every detail of the Gothic Chapel. Bearing as it does, a striking resemblance to the old cathedrals of Europe, the illumination under which it appears tends to the heightening of effect. Nature has shown, her handiwork to advantage here. There is an apparent design in the tout ensemble of the Chapel, a nicety in the separate elaborations which seems the result of a long study of and an intimate acquaintance with the arts. The pillars are so massive and spring towards their proper arches so majesti cally; the tracery is so delicate; and, altogether, there is so harmonious a subordination of one part to another, that we are perfectly confounded at the thought of no human hand having been employed there. Nature, whose common function is to supply the material for human skill to work upon, has acted a double part here, for she has not only given the means, but has blended them into recognized form and proportion. The place is well named, truly. It is religious in every aspect, and the light thrown on it seems heavenly, a beaming from above, to warrant acceptance of the services there performed.

The next object of interest at which the visitor arrives, is called the Devil's Arm Chair. This consists of a large stalagmite column, in the centre of which is formed a spacious seat, as though for the express purpose of affording to his Satanic Majesty all due comfort when fatigued. When the tourists are shown this chair, it is obvious that the same feelings sway them which sway their fellows (of whom we have had some experience) on the occasion of their being shown the Coronation Chair in Westminster Abbey. They are all desirous for the honor of a seat, even though but for a moment, and considerable is the anxiety manifested, lest a chance should be lost.

The Gothic Chapel

More stalactites and stalagmites are passed on the visitor's departure from the Devil's Chair, and then they come, in succession, to Napoleon's Breastwork, the Elephant's Head, and the Lover's Leap. The last named is a large pointed rock, more than thirty feet above the floor, projecting into a large rotunda. It is really amusing to notice the excitement, especially among the ladies, which the mere mention of the name, is sure to upcall. Why the fairer portion of the creation should feel so interested on the subject, and so eager to have the reason of the name explained, can be easily accounted for. They imagine for the moment, that a lover has really sought the shades from that eminence, through either lost or slighted affection; and perhaps more than one fair daughter of Eve will ask whether her true knight would take such a leap for her sake.

Lover's Leap

Immediately below the Lover's Leap is a hollow; turning thence, to the left, and at a right angle to the previous course, will be found a passage or chasm in the rock, three feet wide, and fifty feet deep. This conducts the visitor to the lower branch of the Gothic Avenue, at the entrance to which is an immense flat rock called Gatewood's Dining Table. To the right of this is a cave which is frequently penetrated as far as the Cooling Tub. This is a beautiful basin of water, six feet wide, and three deep. Into this basin a small stream of the forest water pours itself from the ceiling, ultimately finding its way into the Flint Pit, which is at no great distance. Returning thence, the visitor will pass for a second time Gatewood's Dining Table. It will be found to nearly block up the entire way.

Continuing their walk along the lower branch, more than half a mile, and passing, in succession, Napoleon's Dome, the Cinder Banks, the Crystal Pool, the Salts Cave, etc.; and then descending a few feet, and leaving the cave which continues onwards, the visitor enters, on his right, a place of great seclusion, and remarkably grand. It is called Annetti's Dome. A waterfall will be seen through a crevice in the wall of the dome. The water is a foot in diameter, and issues from a high cave in the side of the dome, falling on the solid bottom, and passing off by a small channel into the cistern, directly on the pathway of the cave. The cistern itself is a very large pit, and is usually kept almost full of water.

A beautiful sound like that of distant music will be heard near the end of this branch (the lower). There is a crevice in the ceiling over the last spring; and it is through this, that the ear is cognizant of the falling of water in a cave over head.

Reentering the Main Cave or Grand Avenue, the tourist will discover fresh objects on which to lavish his feelings as well as expressions of admiration, for the might and majesty of the Deity is manifested in many forms and ways. Not far from the stairs leading down from the Gothic Avenue into the Main Cave, will be seen the Ball Room. It has been so named from its manifest availability to the purposes of dancing. It contains a natural orchestra, between fifteen and eighteen feet high, and fully capable of accommodating upwards of a hundred musicians. It also contains a gallery, which extends back to the level of the high embankment near the Gothic Avenue. The avenue here is high, straight, wide, and quite level for several hundred feet. The expenditure of a trifling sum of money would furnish the place with a floor, seats, and lamps necessary for a ball room. Here, the effect of music would be similar to that produced by the same agency in the church; while the scene realized by a ball would be as striking, though certainly less solemn. It would be a truly interesting sight, that splendid avenue, filled by thousands in the various costumes of the world's inhabitants: some engaged in the graceful measure of the dance; others in groups, either looking on, or awaiting their turn to join in the pastime.

The road now before the visitor is a broad and handsome one, although in some places rather dusty. Willie's Spring is soon reached. This locality of the Cave is a beautifully fluted niche in the left hand wall, made by the continued trickling of water into a basin below. The Spring itself has its name from a young clergyman in Cincinnati. This gentleman, having a somewhat romantic temperament, assumed the name of "Wandering Willie." Taking with him his violin, he marched on foot to the Cave. Arrived there, and being fatigued, he selected the spot since named after him, for his place of rest, asking the guide to call him in the morning. But a short distance beyond the Spring, and close to the left wall, is the place where the oxen were fed during the time of the miners; and scattered around are a vast number of corncobs, all quite sound, although they must have remained where they are upwards of thirty years. Near at hand in the wall on the left, will be found a niche. It is of considerable size, reaching from the roof to the bottom of a pit more than thirty feet deep. Down its sides water of the purest kind is constantly dripping. Subsequently it is conducted to a large trough, from which the invalids obtain their required water while they remain inhabitants of the Cave. This well near the bottom expands into a large room, but there is no opening out of it.

The visitor will now pass the Well Cave, Rocky Cave, etc., and arrive at the Giant's Coffin. This is a large rock on the right, and derives its singular name from its obvious resemblance in form to a coffin. It is an object of great interest, and cannot easily be passed by without commanding the notice of even the most careless of tourists.

In this portion of the Mammoth Cave begin those incrustations which have struck so many thousands with wonder and admiration. How varied are they in form; and, separately, how changeable. Yonder you recognize what appears to be the frame of some huge animal. Look again, and you fancy a group of birds are before you; viewed from another point, and some of the finny tribe may be suggested to you. All is wonderful, indeed.

Giant's Coffin.

Proceeding about a hundred yards beyond the is "Coffin," the visitor will perceive that the Cave takes a curve, sweeping round the Great Bend or Acute Angle, and then continuing its proper course. On entering this vast and magnificent amphitheatre, the guide always ignites a Bengal light. The effect is brilliant, such, indeed, as a poet may readily imagine, but with difficulty describe, but of which mere ordinary mortals can have no conception. Enchantment is the only word we can apply to the scene thus presented. As in other portions of the Cave which, like the present, exhibit characteristics of the beautiful or the sublime, we have seen, in the place we mention, the various ways in which a single feeling or passion may be manifested. Delight has been in the ascendant with the visitors, when gazing on the brilliant scene before them: well have they proved its existence! Some by loud expressions, others by silence; some, too, have wept, and many have prayed. How sublime must be the scene to elicit all this!

The Sick Room Cave is opposite to the Great Bend. It bears the name from the sudden sickness of a visitor some years since, it is supposed, from his having smoked, with others, cigars in one of its most confined nooks. A row of cabins, built for consumptive patients, will be found immediately beyond the Great Bend. Two are of stone, the rest are frame buildings. Standing in a line from thirty to one hundred feet apart, their appearance is picturesque, although the visitor cannot divest his mind of melancholy impressions while contemplating them; for the malady which brings so many persons to those habitations is one that attacks the young, that withers beauty and destroys life in its early phase. These houses for the ailing are well furnished; and though we believe that there is no cure for confirmed decline, we are confident that it is to be eradicated in its young or mild stages. The air of the Cave is certainly favorable to such an issue. So convinced of this fact was a physician of high repute, formerly a member of Congress from the district adjoining the Cave, that he was induced to openly express his opinion, that the State of Kentucky ought to purchase it, that it might be enabled to establish a hospital in one of the avenues.

The following remarks on the subject of the Cave's atmosphere, from the pen of a literary gentleman who visited the place in 1832, will not, we hope, be uninteresting to the reader.

"It (the atmosphere) is always temperate. Its purity, judging from its effects upon the lungs, and from other circumstances, is remarkable, though, in what its purity consists, I know not. But be its composition what it may, it is certain its effects upon the spirits and bodily powers of visitors, are extremely exhilarating, and that it is not less salubrious than enlivening. The nitre diggers were a famously healthy set of men. It was a common and humane practice to employ laborers of enfeebled constitutions, who were soon restored to health and strength, though kept at constant labor; and more joyous, merry fellows were never seen. The oxen, of which several were kept, day and night, in the Cave, hauling the nitrous earth, were after a month or two of toil, in as fine a condition for the shambles, as if fattened in the stall. The ordinary visitor, though rambling a dozen hours or more, over paths of the roughest and most difficult kind, is seldom conscious of fatigue, until he returns to the upper air; and there it seems to him, at least in the summer season, that he has exchanged the atmosphere of Paradise for that of a charnel warmed by steam, all without is so heavy, so dark, so dead, so mephitic. Awe and apprehension, if they have been felt, soon yield to the delicious air of the Cave; and after a time a certain jocund feeling is found mingled with the deepest impressions of sublimity, which there are so many objects to awaken. I recommend all broken hearted lovers and dyspeptic dandies, to carry their complaints to the Mammoth Cave, where they will undoubtedly find themselves "translated" into very happy and blithesome persons "before they are aware of it."

The Star Chamber is considered by all visitors to be one of the greatest objects of curiosity in the Cave. It is a magnificent long hall, with perpendicular arches on either side, and a flat ceiling; the side rocks are of a light color, and stand out in relief against the dark ceiling, which is studded with innumerable sparkling substances, resembling stars. The guide on approaching the chamber, takes the lanterns from each visitor, and places them in a hole in the rock, to subdue the light and make the illusion more perfect. Visitors are always lost in admiration, and quit this part of the Cave most unwillingly. The side rocks do not reach within three feet of the ceiling, and no connexion can be seen between the ceiling and the sides, the contrast between the dark ceiling and the light side rocks is so great, that the ceiling appears to be at an immense distance, and after looking at it a few minutes, the visitor fancies he is standing under the canopy of Heaven.

The Star Chamber

Leaving the Star Chamber, the visitor will perceive, in a a kind of cavity in the wall on the right, an oak pole about ten feet long and six inches in diameter, with two round sticks, of half the thickness, and in length three feet, tied to it transversely, at about four feet apart. An ascent to this cavity is made by means of a ladder; the visitor then finds the pole to be firmly fixed, one of its ends resting on the bottom of the cavity, the other reaching across and forced into a crevice about three feet above. The general supposition is, that this was a ladder used by the former denizens of the place, in their procurement of the salts which are incrusted on the walls in several parts. A different opinion, however, was entertained by an intelligent medical gentle man, (Dr. Locke, of the Medical College of Ohio,) that a dead body had once been placed upon it. The Dr.'s reason for this conclusion was based on the fact, that precisely similar contrivances were resorted to by some tribes of Indians, in the disposal of their dead. Strange to say, that though many thousands of persons had explored the Mammoth Cave previous to 1841, it was not until that date that the pole was discovered. Probably, ages as many as we can count from the birth of Christ, have passed over the world since it was placed here; yet it is perfectly fresh, decay has not even marked the bark which confines the transverse pieces.

Some side cuts are next passed through. These, as their names will imply, are caves opening on the sides of the several avenues, and after continuing for some way, rejoining them. Generally speaking, they are short, though some of them are more than half a mile in length. Quartz, chalcedony, red ochre, gypsum, and salts are found here. Slowly the visitor wends his or her way to the Salts Room, the ceiling and walls of which are covered with salts hanging in crystals. These frequently fall like flakes of snow, through the agitation of the air. In this Room will be seen the Indian Houses. They are under the rocks, and are small rooms entirely covered. Many of them contain cane partly burnt, and ashes.

The next portion of the Cave visited is the Cross Room. It is a principal section of the avenue. The most extraordinary feature of the Cross Room is, that it has a ceiling of one hundred and seventy feet span, and yet not a single column to support it.

The portion of the Cave immediately succeeding the place we have been describing, is known by the ominous title, Black Chambers. They contain several ruins, which consist of large blocks of different kinds of strata. They are cemented together, and bear a striking resemblance to the walls and pedestals of the old baronial castles of European countries. Here the avenue is very wide, sufficient to render it tedious to walk from one side to the other. A little way beyond the ruins, and on the right, the tourist will enter the right branch. It is on the same level, and the ceiling is regularly arched. An ascent is then made through the

Big Chimneys, into an upper room. This is about the same size as the Main Cave, the bottom of which is somewhat higher than the ceiling of the one underneath. The low, plaintive murmurings of a distant waterfall are heard here as the visitor proceeds. They grow louder and louder, until we find ourselves close to the Cataract, when our ears are cognizant of a perfect roar. Very large perforations are seen in the roof, on the right hand side, from which water is always falling, generally not in considerable quantities, but after heavy rains, in complete torrents, and with a roar that resounds afar, and seems to be shaking the Cave itself from its very foundations. The water falls into a funnel shaped pit, and is lost to the sight in less than a second.

Parties occasionally select this portion of the Cave as their dining room. Their previous explorations have given them a keen appetite, which, from experience already gained, they know cannot be better satisfied than by the excellent fare of the Cave House; they have been careful to furnish themselves from the abundant larder of that establishment, and after the arrangement of the plates, decanters, etc., prepare to do the "honors" with a gusto that none but hungry people can know or appreciate. The incidentals of a banquet like that we refer to, are necessarily very interesting. The time, the place, and the varied characteristics of the visitors all tend to the perfection of enjoyment. We have noticed how different dispositions have become harmonized, how, too, the good wish of Macbeth towards his guests, has been fulfilled or realized in this Mammoth Cave; certainly nothing could be more perfect than the "digestion that waited on appetite."

From the particular point of the Cave to which we have brought the visitor, a view of the main avenue will be caught on the left. Continuing its general course, it exhibits the same features of grandeur and solemnity as were noticeable at the first. We have also pointed out to us the way to the Humble Chute and the Cataract. The Humble Chute is the entrance to the Solitary Chambers; but before entering it we find it necessary to crawl on our hands and knees, between fifteen and eighteen feet under a low arch. The term "solitary," as applied to these chambers, is exceedingly appropriate; a person once in them, feels himself entirely secluded from his fellow men. The profound sense of loveliness will grow upon him, until, indeed, it becomes almost too powerful for endurance. In this Cave is a Fairy Grotto, a place that is a perfect realization of its name, it is fanciful and pretty: just such a place as one of the it "little people" would select as a habitation. Stalactites by the tens of thousands are seen, at various distances, extending from the roof to the floor. They are of various sizes, and of shapes that, in their variety at least, can only be equalled by the kaleidoscope. They form columns, irregularly fluted, and others quite solid. These are near the ceiling, and, divided lower down, into small branches similar to the roots of trees, exhibit the appearance of a coral grove. The lamps with which the visitors are furnished, being placed close to the incrustations on the columns, this forest of stalactites is thoroughly lighted up, and discloses a scene of the rarest and most astonishing beauty. But lovely as this portion of the Cave is, it was much more so a few years since, before numerous disgusting caricatures of civilized creatures broke away many of the stalactites; and, as a memorial of their ignorance and brutality, left them strewn in wild confusion on the ground.

Entering the main Cave at the Cataract, from the Fairy Grotto, the walk is continued to the Chief City or Temple. This is an immense vault. It extends over an area of two acres, and is crowned by a dome of solid rock, one hundred and twenty feet high. Perhaps equal in size to the Cave of Staffs, it may be considered as a rival to the far famed vault in the Grotto of Antiparos, which has the repute of being one of the largest in the world. There is a remarkable circumstance connected with this portion of the Cave, which cannot fail to attract the notice of even the dullest tourist. It is the moving appearance of the dome, as the visitors pass along, suggestive of the same appearance in the sky, when the pedestrian is above the surface of the earth. A large mound of rock rises in the middle of the dome, nearly to the top. It is very steep, and forms what is called the Mountain. Many persons, when ascending this mound from the Cave, experience feelings of awe more intense than they ever knew before. We can well believe such to be the case. Immediately around the tourist is a narrow circle, illuminated; without it, both above and beyond, there seems to be unlimited space. No object is there for the eye to fasten on, no sound, breaking the dull monotony, for the ear to catch. A vague fear creeps upon the visitor, although he knows he is bounded by stone walls, and actually safe. But it is necessary to penetrate that profound darkness, think the Cave's frequenters, and accordingly, numerous fires are kindled. So gradually as they flicker upwards, so gradually is presented to the sight a spectacle of transcendent magnificence. Let the reader fancy a strata of gray limestone, opposite him, breaking into steps from the bottom to the top, scarcely discernible; above, the lofty dome, having at the top, a smooth slate of oval form, in outline exquisitely defined; and the walls sloping away from it till lost in the darkness of night. A well known writer on the Mammoth Cave, and its frequent visitor besides, (Lee) thus concludes some remarks he makes upon the temple:

"Every one has heard of the dome of the Mosque of St. Sophia, of St. Peter's, and of St. Paul's; they are never spoken of but in terms of admiration, as the chief works of architecture, and among the noblest and most stupendous examples of what man can do when aided by science; and yet, when compared with the dome of this Temple, they sink into comparative insignificance. Such is the surpassing grandeur of Nature's works."

At the back of the Giant's Coffin is a narrow passage leading into a circular room, of about one hundred feet in diameter, with a low roof. It is designated the Wooden Bowl, some say from the resemblance it bears to one; while according to the account of others, the name is derived from the circumstance of a bona fide wooden bowl having been found in this portion of the Cave, by a miner. On the right the visitor will perceive the Steeps of Time, descending which for about twenty feet, he enters the Deserted Chambers. Their characteristics are very wild and varied. Advancing two hundred yards, your notice will be attracted by the ceiling, which presents a rough and broken appearance for a little while, and then shows a surface waving, white, and smooth. At Richardson's Spring we distinguish, as we have already done in other portions of the Mammoth Cave, the tokens of a bygone age and people. They are the imprints of moccasins and of children's feet.

The Bottomless Pit

The pits in the Deserted Chambers are numerous, more so than in any other portion of the Cave: the Covered Pit, the Side Saddle Pit, and the Bottomless Pit are the most noted. The entire range of these Chambers is, in fact, so alternated with pits, and is so bewildering from the serpentine form and irregularity of its branches, that the visitor is not over anxious to roam far from his guide, who, of course, is intimately acquainted with every rood of ground to be covered.

The Covered Pit is in a little branch to the left. It is between twelve and fifteen feet in diameter. A thin rock covers it, having a narrow crevice, with only a trifling support on one of the sides. A large rock rests on the centre of the cover, and the sound of a waterfall may be heard, though the fall itself cannot be seen.

The pit next in succession, the Side Saddle Pit, is about twenty feet long and eight feet wide. It has a margin about three feet high, extending lengthways ten feet. Against this, the visitor may lean, and view the interior of both dome and pit.

The place we now come to is the Labyrinth, entrance to which is gained, after a short walk from the Side Saddle Pit, by means of a ladder. We make a descent. One end of the Labyrinth leads to the Bottomless Pit, which it enters about fifty feet down; the other, after taking numerous and fantastic windings, conducts to Goran's Dome. This, though of recent discovery, is one of the most attractive features of the Cave.

Louisa's Dome is in the neighborhood. It is an exceedingly pretty little place, not more than twelve feet in diameter, but about twenty four in height. Immediately under the centre of the cave just left, and lighted up as it frequently is, the visitors in it can be seen distinctly from above, through a slight crevice in the rock.

In Goran's Dome it is impossible not to be struck with the apparent intention of producing certain effects, which is a characteristic of the place. We have noticed the same kind of thing as manifest in another portion of the Cave: the Gothic Chapel. Goran's Dome seems to have been specially made for a certain purpose; in every way it is adapted to the convenience of numerous visitors.

At the termination of the Labyrinth, we come to a kind of natural window, about four feet square, and three feet above the floor. This window opens into the interior of the Dome, about equidistant from the top and bottom. The wall of rock here is of trifling thickness, not more than eighteen inches; it continues around on the outside of the Dome, and along a gallery of a few feet in width, for upwards of twenty paces. A good view of almost the entire space within is gained at another opening of much larger size. The Dome is of solid rock, and very lofty, perhaps two hundred feet; its sides seem fluted and elaborately polished. One of the wonders contained in this portion of the Cave is a huge rock, about thirty feet from the window, and directly fronting it. Its resemblance to a curtain is very striking; we fancy we distinguish folds in it, and look upwards in the endeavor to ascertain to what it is suspended.

It may be necessary here to state, in order to account more satisfactorily for what we shall have to say anon, that it is customary for the guides to leave those on whom they attend when the latter are expressing their admiration of this curtain and the details in its vicinity. With feelings wound up to a pitch of ecstacy by the things before them, they have no idea of the change which awaits them. Suddenly as thought, and as we may imagine light came at the behest of God, the whole place is illuminated. Every thing above, below, or around is shining in brilliant light. The sides of the fluted walls, the yawning gulf, the first towering to an indefinite height, the other descending to where it would seem no soundings could be gained. By what means the sudden illumination has been caused, we need not say, for we have already mentioned the sudden exit of the guides.

In leaving Goran's Dome, an ascent is made by means of a ladder near Louisa's Dome; and we continue on, having the Labyrinth on our right, until it is lost in the Bottomless Pit, which also terminates the range of the Deserted Chambers. These were once considered as the end of all exploration; and perhaps this mistaken notion would have continued to this day, had not a Mr. Stephenson of Georgetown, Kentucky, and a well known and adventurous guide named Stephen, presumed that there were things to be seen on the opposite side in every way as attractive as the objects with which tourists were already familiar. The means hit upon were a ladder. This idea was soon acted on, and a stout ladder was eventually thrown over a chasm of twenty feet wide, and more than two hundred feet deep. The Bottomless Pit, similar to a horseshoe in form, has a tongue of land twenty seven feet long, and runs out into the middle of it. From the extremity of this land, a bridge of substantial workmanship, has been thrown across to the cave on the opposite side. Some idea of the depth of the pit, and of the ideas of awe which a contemplation of it awakens, may be imagined, when we state that it is the practice of the guides to let down pieces of lighted paper into the abyss, which, descending lower and lower, ultimately vanish from the vision, though long before they are extinguished.

Having crossed the bridge to the opposite Cave, the visitor finds himself in the midst of a rocky world. Over the pit a dome of magnificent dimensions expands itself. We then proceed along a narrow passage for a little distance, and reach a point, whence diverge two routes, the Winding Way and Pensico Avenue.

The former is about fifty feet in width, with a height of about thirty feet. Different estimates have been given of its length, but taking the mean, we should be inclined to say it was about two miles. The beautiful and the sublime are mingled here; and dull, indeed, must he be who is not open to their influences. It is most interesting throughout. For upwards of a quarter of a mile from the entrance, the roof is exquisitely arched, about twelve feet high and sixty feet broad. Formerly, it was encrusted with rosettes and various other formations. They have been either taken away or demolished; and thus another instance is afforded of the wretched vandalism which people calling themselves civilized, sometimes indulge in. Here the walking is very good, and so wide, that a dozen persons could run or walk abreast for a quarter of a mile, to Bunyan's Way. This is a branch of the Avenue leading on to the river.

The Avenue now changes its features. They have been those of beauty and regularity; they are now of grandeur and sublimity, and are continued to the end. The way is no longer even and smooth; on the contrary, huge rocks are scattered about in the wildest disorder. The roof, lofty and magnificent, presents its long, pointed, or lancet arches, suggesting .a strong likeness to the elaborate ceilings of the ancient Gothic Cathedrals of Europe. Feelings of true religion must necessarily be upcalled here. The forms and ceremonials of faith respectively acknowledged by the visitors may be, and most likely are different, but that feeling of devotion and love, which are necessarily at the foundation of all religious opinions, are there; it is the same God who is worshipped,

"Jehovah, Jove, or Lord!"

Making a gradual descent of a few feet, we enter a tunnel of fifteen feet wide. It is on the left, and the ceiling, twelve or fourteen feet high, is perfectly arched and beautifully covered with white incrustations. Hence the Great Crossings are very soon gained. In this part of the Cave, it may be remarked, the guides usually jump down some six or eight feet, from the place previously examined. Where we now stand is a tunnel, and following our guide's example, we soon rejoin him. The name Great Crossings is by no means an unexpressive one; though we must clear the mind of an erroneous opinion held regarding it, namely, that it was given in honor of the Great Crossings, where resides the man who killed Tecumseh. The name was really adopted because two large caves cross here.

No great distance from this point, an ascent is made up a hill on the left. Proceeding a little way, over the ancles in dry nitrous earth, somewhat at a right angle with the avenue below, the visitor arrives at the Pineapple bush. Why this locality bears so euphonious a designation, it would be difficult to tell. It is secluded in a very pretty spot; the air is fresh and reviving, and beneath its influence the blood dances merrily through the veins. Indeed, we would recommend the hypochondriac and love sick, as well as the admirers of romantic scenery, to visit this place.

Descending into the avenue already explored, a great number of stalactites and stalagmites are passed in succession. These bear a strong resemblance to coral. In the vicinity are to be found a multitude of crystals of dogtooth spar, shining most brilliantly. The place is called Angelica's Grotto; and so beautiful is it as to impress its own sanctity on every person with a mind susceptible of high influences; yet did a clergyman systematically destroy a number of crystals, that he might inscribe his name in the vacancy thus left; and what gave a spirit of cowardice to the act, this reverend gentleman watched till the back of the guide was turned towards him, ere he attempted the act of barbarity. Had he done it boldly, there would have been some mitigation.

The visitors will now return to the head of the Penisco Avenue, then turn to the right, and enter the narrow pass which leads to the river, following its course for some hundred yards; descending all the while down a ladder, or stone steps. Finally, they come to a path, which is cut through a high and broad embankment of sand. This conducts to the far famed winding way. In the opinion of many persons, whose conclusions are entitled to deference, this way has been channeled in the rock by the gradual attrition of water. Should this be the case, and there is certainly nothing to controvert the hypothesis, the imagination is lost in finding a date for the commencement of the work. Was it contemporaneous with the birth of Christ or with the reign of Solomon, or, going to more remote times, the era of Moses? Beyond, far beyond them all, to a day many more thousands of years before us than, according to written tradition, the world itself has seen.

In length, the Winding Way is one hundred and five feet, in width eighteen inches, and in depth seven feet. It widens out above, sufficiently to permit the free use of a man's arms. Throughout, it is a zig-zag, and the reader may imagine that those persons who are of very corpulent proportions, do not like very much to pass through it. No companies of visitors can possibly enter this passage, except in single file, and it is quite amusing to see how ladies and gentlemen of the most delicate proportions eye the sides askance, to satisfy themselves that a squeeze is not in reserve for them.

Aptly enough, the place entered after quitting this limbo is called Relief Hall. It is very wide and lofty, but not long. Taking the turn to the right, the termination at River Hall, distant about one hundred yards, is soon gained. Two routes will now be visible, that on the left, conducting to the Dead Sea and the Rivers, that on the right, to the Bacon Chamber, the Bandit's Hall, the Mammoth Dome, besides a variety of other caves, domes, etc.

The Bacon Chamber, like Relief Hall, is very aptly named. It has a low ceiling, clustered with canvassed hams and shoulders. The visitor will next proceed to the Bandit's Chamber, where he will arrive after a steep ascent of twenty or thirty feet. Huge rocks obstruct the way; the consequence is that the passage is extremely difficult, and that tourists are obliged to clamber on the best way they can. There is a striking similitude between the name and the aspect of the place It is a chamber vast and lofty. The floor is covered with a mountainous heap of rocks, which rise in an amphitheatrical form almost to the ceiling, and are so arranged as to furnish, at different degrees of elevation, galleries or platforms, reaching immediately round the chamber itself, or leading off into some of its hidden or secret recesses.

Unexpectedly as before, and in another portion of the Mammoth Cave, this Hall is suddenly lighted up. The guide has just slipped away from the company, to ignite his torch, and now the awful looking roof, the towering cliffs, and the clustering rocks are bathed in a brilliant glare. It is almost out of the power of words or of the artist's pencil to portray the scene presented. Such a spectacle must be witnessed by the eye to be at all truly understood.

Diverging from the Bandit's Hall are two caves; that on the left, leads to a multitude of other domes; the other, on the right, is called the Mammoth Dome. Directing his steps towards the latter, the visitor, after a rough walk of nearly a mile, comes to a platform commanding a somewhat indistinct view of this magnificent dome. It is of recent discovery, not more than two years having elapsed since it was first explored by a German gentleman and the guide Stephen. From the platform, persons are sometimes let down by ropes, a distance of twenty feet, and when on the ground, find themselves on the side of a hill, directly under the great dome, from the summit of which there is a water fall. This, dome is of immense height, more than four hundred feet, and is considered, with perfect justice, one of the most magnificent objects in the world. The visitors who have been lowered, as we have described, ascend the hill to which they were previously let down from the platform; continuing thence up a remarkably steep hill, for more than a hundred feet, they reach its summit. A scene of awful sublimity is commanded here. Looking down the declivity, the visitor will perceive, some distance to the left, those other visitors whom he has just quitted. The latter are standing on the platform or termination of the avenue along which they have come; lower down yet, the bottom of the great dome itself presents itself. Overhead, at a distance of more than two hundred and eighty feet, the ceiling is lost in the deep obscurity. It may be necessary for us to state that the height of the ceiling was determined by Mr. E. F. Lee, civil engineer, and subsequently confirmed by discovering on the summit of a hill, where man's foot had never trodden before, an iron lamp. The astonishment of the guides on the occasion may be better imagined than described. They, as well as the visitors with them, were fairly astounded; and doubt respecting the cause would have continued down to the present date, had it not been for the following circumstance: An old man was living at the Cave Hotel. He had been employed as a miner in the saltpetre establishment of the Messrs. Wilkins and Gratz, some thirty years previous; and on being shown the lamp, he said without the least hesitation, that it had been found under the crevice pit. The name surprised all present; but the old man continued to inform his questioners, that during the time his late employers were engaged in the saltpetre trade, a person of the name of Gatewood stated to Wilkins, that the probabilities were, that the richest nitre earth was under the crevice pit. Its depth was, at that day, totally unknown, and Wilkins, to ascertain it, took a rope 45 feet long, and fastened this very lamp to it. He then caused it to be lowered into the pit, but while descending, the string accidentally caught on fire; the result was the loss of the lamp. Wilkins, however, made an offer of two dollars award for its recovery. It was accepted by a miner, who, on account of his diminutive stature, was nicknamed Little Dave. With a rope round his waist, and torch in hand, he was lowered to the depth of forty five feet, and then drawn up. On regaining the spot from which he had descended, this miner was found to be dreadfully alarmed. He trembled in every limb, his eyes were fairly starting out of his head, and not a single word could we get from him for the first five minutes. Subsequently, however, he began to collect himself, and said that no amount of money, not even the National Debt of England, would induce him to make the trial a second time. By way of excusing this resolve, expressed so firmly, and to be kept so religiously, Little Dave entered into the most astounding relations. Baron Munchausen's fictions were probabilities compared to what our miner related; but it is supposed that the height at which poor Dave was suspended, upwards of two hundred and forty feet, was the only course of his alarm. Like many other gentlemen, he was not constituted for an exalted station, and the accident that placed him in it, though only for a short time, took away all the sense with which nature had gifted him. But the correctness of the old miner's estimate was established, guides being sent to the place where the lamp was found, other persons being stationed at the same time at the mouth of the crevice pit.

Words were exchanged by the parties, and sticks thrown; and by these means the truth was made known.

From the mouth of the cave to this pit the distance is not very long, less than half a mile; notwithstanding, a circuit of more than three miles must be made, before the grand apartments immediately beneath, can be reached. The illumination cast upon that portion of the great dome on the left, and of the Hall on the top of the hill bearing to the right, as witnessed from the platform, is certainly one of the most sublime spectacles we can call to mind, although our experience in such matters is considerable. To be seen to the greatest advantage, however, a different position should be taken by the visitor. Then should the Bengal lights be ignited, and by their aid, the dome, apparently towering to a neighborhood with the heavens; the hall on the summit of the hill, with its multitude of stalagmite columns, all so majestic; and its other details, each beautiful in itself, and the whole harmonious in their arrangement, will be seen.

That the exclamations of admiration elicited by a sight like this, are many and correct, who can doubt Europeans, who are good judges of what is beautiful or sublime in Nature, have been heard to declare, that to see such a sight as that presented by the dome under illumination, they would gladly travel ten thousand miles. Several liable to illness, and whose voyage across the Atlantic, speedily and comfortably as it is now performed, has entailed much sickness, have confessed that the spectacle is a perfect compensation.

With this portion of the Mammoth Cave concludes, in most cases, the day's explorations. True, that the sun has not set, and that there is plenty of time left; but, no! the visitors feel anxious to return. How is this? Can we account for it? Yes; the tourist's feelings are a realization of what Byron has recorded on his immortal page. With perfect propriety can it be said of the admiring company, that

"Dazzl'd and drunk with beauty,"

they are in no fit mood to seek for other objects of interest. They have seen enough, and they now come from the great, the sublime, the crowning one, deeply penetrated by its effulgence.

Another scene of wondrous sublimity awaits the visitor in the morning. He has heard from others of it, and begins to wonder whether their accounts have not been over colored; whether, in fact, what he is about to see will come up to, or can surpass, what he has already seen. He thinks not, and prepares to ascertain. New from the slumbers of the night, and refreshed by both the pure air and the wholesome viands of the Cave House, he starts again with more like himself and the faithful guide.

With a pace accelerated by expectation, the party proceed onwards, and arriving at the Cave, seek the River Hall. It is there before them, sublime in its every feature. The river has been recently overflown. That is manifest, for how flush, how heaving it is.

The River Hall descends similar to the slope of a mountain; and like the firmament at midnight, when the stars come forth in their glory, the ceiling stretches away, away, it seems to infinity. You proceed onwards, making a gradual ascent, and keeping pretty close to the right hand wall. You will then observe on the left, a steep precipice. Over this you will look down, being able to do so by numerous blazing missiles, upon a broad black sheet of water, eighty feet below. It is called the Dead Sea; and the name, so awful and so referrable to awful events, cannot be better verified than here. There is a terrible grandeur in the place. Long after you have left it, the eye continues cognizant of its many sights, the ear of its many sounds. The memory holds them, and they even haunt the dreams of night.

The descent is made by means of a ladder. It is of about twenty feet; the visitors then find themselves in the midst of gigantic rocks, heaped pile upon pile. In the mingling of lights and shadows, the persons who have come to see the River, although dressed in modern fashionable style, will seem of the locality a fitting race. Slowly they move in files, men, women, and children being together, with lamps in their hands. These lamps are guarded with extreme care, as they are liable to go out through any in advertence. Gradually their illumination falls upon the different details; the ceiling,the walls, the cliffs, the ravines. Now the light, thrown upwards, is reflected through the fissures in the rocks; presently it is reflected from towering cliffs, every outline of which it defines, thus relieving the most intense darkness beyond. In some parts the water is not seen, although it is heard; but its murmur sounds. awfully. In others, its appearance is brilliant through the light of the lamps.

At the foot of the slope the River Styx winds its way. It is aptly named: people might well imagine it to be the fabled stream whose name it bears. Four passengers only can be conveyed over this river at the same time. The guide fastens lamps at the prow of the boat, and the various images are reflected in the murky pool.

There is another mode of crossing the Styx. It is by means of a bridge overhead, composed of abrupt precipices. To avail himself of this bridge, the tourist must ascend a very steep cliff; then enter a cave above, three hundred yards long. Leaving this, he will find himself on the bank of the River, more than eighty feet above its surface. He will then command a view of the persons who are in the boat, and also of those upon the shore. The lamps in the canoe, when viewed from this distance, have a singular and striking appearance. Their glare is that of gigantic eye balls. Sitting somewhat in the shade, the mere outline of the visitors' figures can be seen, and they look like so many shadows, the spirits of the departed, being rowed over that profound flood to a place where final doom is to be awaited.

Turning their eyes from the boat and its contents, the persons on the shore will see those of their companions who, like themselves, have come over the bridge, grouped very fancifully.

74

Their appearance is much less spectral than that of the people on the water. They seem human still, and give a warranty that a return to the upper world is possible.

The Styx is the smallest river in the Mammoth Cave. Having passed it, the visitor walks over a pile of large rocks, and finds himself on the banks of the Lethe. Here, again, will be found a striking resemblance between natural objects and the names given them. How striking is forgetfulness typified in that river! We remember seeing many years ago a picture of the Waters of Oblivion, painted by John Martin, which, in its general details, in the *tout ensemble*, might have been taken as a representation of this cave stream and the objects which surround it.

The River Styx

Looking back, the tourist will perceive a line of men and women descending the high hill from the cave, which runs over the river Styx. Two boats are kept, and the parties who have come by the two routes, that is, either down or over the Styx, may unite and descend the Lethe about a quarter of a mile. Throughout the whole distance the ceiling is very high, upwards of fifty feet we should say. On landing, a lofty and level hall is entered. It is called the Great Walk, and extends to the banks of the Echo, a distance of three or four hundred yards. The Echo is a bona fide river, wide and deep enough, we believe, to float a steamship as large as the "Atlantic" or "Pacific." On the occasion of embarking, persons will readily notice the lowness of the arch: three feet constitute the entire passage left for the boats. The consequence of this is, of course, a complete doubling up on the part of the passengers. In deed, their position is anything but comfortable while they are being rowed over, but their sufferings, if we may so designate their sensations, are not of much duration, a couple of boats' lengths being quite sufficient to put them where no complaint can be made of the Cave, so far as the advantages of height and width are regarded. The boats used here are capable of carrying twelve persons each. The passage down the river is replete with pleasure and interest. The extraordinary character of the scene, its magnificence, must necessarily awaken the highest feelings of admiration. A sudden ecstasy is upcalled, which, after holding possession of the soul for a time, subsides, and is succeeded by sensations of quiet felicity. Few persons who ever witnessed the scene, we think, could have allowed an angry feeling to find a dwelling in their bosoms, while under its influences. Powerfully, most powerfully is the benign mandate of Christ, for those whom he redeemed to live in love and peace with one another, impressed here. Nature, in her aspects of beauty, magnificence, and solemnity, is a mighty illustrator of Him whose work she is; and there are thousands of instances of her power to improve or purify those on whom both oral and written precepts have had no power. May we not believe that the stream of Lethe in this Mammoth Cave of Kentucky, had it a voice, could tell us of such changes, wrought on its bosom or its banks, in the souls of many a visitor?

Low and musical is the rippling of the water, as heard by those in the boat as well as on the shore. Beating under low arches and into the cavities of rocks, it may be called the very pulse of the place. So strongly does sound magnify itself in this portion of the cave, that the report of an ordinary pistol is like that of the heaviest artillery. It is prolonged for minutes, and ultimately dies away in low thunder like mutterings.

We remember well on the occasion of a late visit to the Cave, that the whole boat's company joined simultaneously in song while gliding down this river Lethe. How sublime, how truly religious was the effect! Many are the cathedral choirs we have listened to, and deep have been our impressions of their excellence, but they did not amount to those we experienced here. That such was the fact cannot cause surprise. The cathedral, elaborate as it may be in its carving and emblazonry, the beauty of its stained windows, and the skill of the painter, is, after all, no more than a human temple, a pile erected by hands; and its choir, but a band of trained singers. But here, in this Mammoth Cave, we gaze upon the temple of God himself, while the hymnings we listen to are those of spontaneous worship, bursting from the rapt soul, and ascending to the Sanctuary of heaven.

Sometimes a full band of music has been tried on the Echo. What the effect has been can be imagined. Truly may it be said that such things cannot be obliterated from the memory. Let us add, that they ought not to be, for they assuredly make better creatures of us all.

The river Echo is about three miles in length. There is a rise in the water, of only a few feet, through which the three rivers are united. When there has been a long succession of heavy rains, these rivers sometimes rise to a perpendicular height of more than fifty feet, and, with the cataracts, exhibit an aspect of awful grandeur. When the rise of the water does not extend even beyond two feet, the low arch at the entrance of the Echo cannot be reached by the visitor. Occasionally, great apprehensions have been felt by the tourists, in consequence of their being caught on the opposite side, by a sudden rise; but the guide has considerately informed them of an upper cave, admitting of a passage, leading round the arch to the Great Walk.

Purgatory is the name applied to this cave or passage. Once, for a distance of more than forty feet, the visitors were obliged to crawl their way through it, in consequence of its lowness; lately, however, it has been enlarged, and now persons can walk erect, to their entire satisfaction.

Through this improvement, an excursion can be made to Cleveland's Avenue, almost entirely by land, and the tourist will rest satisfied of his not being caught beyond the Echo. In that river and the others which are found in the Mammoth Cave, that very extraordinary fish, the White Eyeless, are to be seen. On the occasion of our last visit to the Cave, we were shown two of them. We, as well as the persons with us, examined the fish attentively, but not one of us was able to distinguish anything like an eye; nor have the skilful anatomists who have experimented upon them, been at all more successful. Indeed, it has been asserted by men most celebrated in their profession, that these fish are not only without eyes, but also exhibit other anomalies in their organization, highly interesting to Naturalists. At the time the rivers of the Mammoth Cave were first crossed (1840), and since, several endeavors were and have been made to discover whence the White Eyeless fish come, and, also, whither they go; but though various conjectures have been formed, nothing that can be looked upon as satisfactory has been arrived at. All is still mystery, and we suppose will continue so until the end of time.

The barometrical measurement of the rivers in the Mammoth Cave has been frequently taken. According to Professor Locke, they are on a level with the Green River. But Mr. Lee, civil engineer, is of a widely different opinion. He says: "The bottom of the Little Bat Room is one hundred and twenty feet below the bed of Green River. The Bottomless Pit is also deeper than the bed of Green River; and so far as a surveyor's level can be relied on, the same may be said of the Cavern Pit and others."

We may remark here, that at the time of Mr. Lee's visit in 1835, the rivers of the Cave were unknown, and that there is no doubt as to their being lower than the bottom of the pits, or of their receiving the water flowing from the latter. If we take the statement of Lee as correct, the bed of these rivers is lower than the bed of Green River at its junction with the Ohio, taking as conclusive the report of the State Engineers, relative to the extent of fall between a point above the Cave and the Ohio, and of the entire correctness of this report we cannot entertain a doubt. Mr. Lee thus continues his remarks on the subject: "It becomes then an object of interesting inquiry to determine in what way the waters are disposed of. If they are emptied into Green River, the Ohio, or the ocean, they must run a great distance under ground, and have a very trifling descent."

We must again join the visitor in his explorations. Leaving the Echo, a walk of four miles brings us to Cleveland's Avenue. Between the river and this point, several objects of interest are met with. Their enumeration would be confusing; therefore, in the trust that all who read these pages will visit the Cave itself, if they have not done so already, we will proceed with our more immediate details.

We pass in succession El Ghor, Silliman's Avenue, and Wellington's Gallery, to the foot of a ladder. This leads up to the Elysium of Mammoth Cave. Those among our friends who are subject to weariness or thirst when walking, will be glad to hear that Carneal's Spring is in the portion of the Cave we are describing; many thousands slake their thirst here. A sulphur spring is also close at hand. Its water equals, both in point of quantity and quality, that of the celebrated White Sulphur Spring of Virginia. Standing at the head of the ladder we have named, the visitor finds himself surrounded by stalactites; they are above as well as about him. With little or no aid of the imagination, these stalactites may be taken for magnificent clusters of grapes. Plump, round, and polished, they would, even on a nearer inspection, do credit to the sculptor's art. The place where they are found is called Mary's Vineyard, and is the commencement of Cleveland's Avenue, one of the chief glories and wonders of this underground world.

About one hundred feet from this spot, taking the right, over a rough and rather difficult way, the tourist at last reaches what is called the height or hill. On this stands the Holy Sepulchre. This natural chapel is about twelve feet square; it has a low ceiling, and is decorated in the most magnificent style imaginable, having well arranged draperies of stalactite of every possible shape. You go on to the room of the Holy Sepulchre adjoining. Unlike the place you have left, it is without ornament or decoration of any kind whatever; it presents nothing but dark and bare walls, and has been likened by many who have been there, to a charnel house. In the centre of this room, which stands but a few feet below the Chapel, the visitor will be shown what seems to be a grave hewn out of the solid rock. So great is the resemblance as to have suggested to a Roman Catholic priest the exclamation which has since passed as its name. The reverend gentleman referred to, no sooner cast his eyes upon this opening in the rock, than he cried out, on bended knees, and with uplifted hands, "The Holy Sepulchre! The Holy Sepulchre!"

This reminds us very forcibly of an occurrence we witnessed some years since, at the annual exhibition in London of the late Haydon's pictures. Among them was a representation of the Crucifixion, so life-like in all its details, and so melancholy true in its chief figure, that of the dead Messiah, that an eminent foreigner, a Catholic, on seeing it, threw himself, it seemed instinctively, on his knees, and kissed the canvas.

Returning from the Holy Sepulchre, the tourist commences his wanderings through Cleveland's Avenue. It is of considerable length, extending from one end to the other, three miles; while its height and width are respectively fifteen and seventy feet. This Avenue is truly magnificent; it may be designated one of the most magnificent objects in the world. It is replete with formations that are to be seen in no other places; which even the dullest cannot behold without experiencing sensations quite new to them, but which, in the cultivated and intellectual, awaken feelings of rapture.

Professor Locke has designated some of these formations as onlophilites, or curled leaf stones; in lecturing on them, he says, "They are unlike anything yet discovered, equally beautiful for the cabinet of the amateur, and interesting to the geological philosopher."

Another gentleman, (a clergyman,) also speaks of these formations. His remarks are to the following effect.

So exquisite and beautiful is Cleveland's Avenue, (the place where they are found) that it is out of the power of painter or poet to conceive anything like it. Such loveliness cannot, indeed, be described. Were the sovereigns of wealthy states to spend their all on the most skillful lapidaries they could find, with the view of rivaling the splendor of this truly regal abode, the attempt would be entirely vain. What then, is left for the narrator? People must see it; and then they will be convinced that all attempts at adequate description are useless. The Cabinet was discovered by Mr. Patten of Louisville, and Mr. Craig of Philadelphia, accompanied by the guide Stephen. It extends in nearly a direct line, about one mile and a half, (some persons say two miles). It is a perfect arch of fifty feet span, and of an average height of ten feet in the centre, just high enough to be viewed at ease in all its parts. It is encrusted from end to end with the most beautiful formations, in every variety of form. The base of the whole is carbonate (sulphate) of lime, in one part of dazzling whiteness, and perfectly smooth; and in other places, crystallized so as to glitter like diamonds in the light. Growing from this, in endless diversified forms, is a substance resembling selenite, translucent and imperfectly laminated. It is most probably sulphate of lime (a gypsum) combined with sulphate of magnesia. Some of the crystals bear a striking resemblance to branches of celery, and all are about the same length, while others, a foot or more in length, have the color and appearance of vanilla cream candy. Others are set in sulphate of lime, in the form of a rose; and others still roll out from the base in forms resembling the ornaments on the capital of a Corinthian column. Some of the incrustations are massive and splendid, others are as delicate as the lily, or as fancy work of shell or wood. Let any person think of traversing an arched way like this for a mile and a half, and all the wonders of the tales of youth, not forgetting those gorgeous fictions, "The Arabian Nights," seem tame and uninteresting, when brought into comparison with the living, growing reality. The term "growing" is not a misnomer; the process is going on before your eyes. Successive coats of these incrustations have been perfected, and then crowded off by others; so that hundreds of tons of these gems lie at your feet, and are crushed as you pass, while the work of restoring the ornaments for Nature's boudoir is proceeding around you. Here and there through the whole extent, you will find openings in the side, into which you may thrust the person, and often stand erect in little grottos, perfectly incrusted with a delicate white substance, reflecting the light from a thousand glittering points. Many visitors are so enraptured with the place, that they cannot repress exclamations of surpise or worship. With general unity of form and appearance, there is considerable variety in the Cabinet. The Snow Ball Room for example, is a secretion of the cave described above, some two hundred feet in length, entirely different from the adjacent parts; its appearance being aptly indicated by its name. If a hundred rude school boys had but an hour before completed their day's sport, by throwing a thousand snow balls against the roof, while an equal number were scattered about the floor, and all petrified, it would have presented precisely such a scene as you witness in this room of Nature's frolic. These "snow balls" are a perfect anomaly among all the extraordinary forms of crystallization. They result, it is supposed, from an unusual combination of the sulphates of lime and magnesia, with a carbonate of the former. We found here and elsewhere in the cabinet, fine specimens of the sulphate of magnesia, (Epsom salts) a foot or two long, and three inches in thickness.

Leaving the quiet and beautiful cabinet, you come suddenly on the Rocky Mountains. Compared to the previous scenes, they furnish a contrast so bold and striking as almost to startle the visitor. Clambering up the rough sides some thirty feet, he passes close under the roof of the cavern he has just left, and finds before him an immense transverse cave, one hundred feet or more from the ceiling to the floor, with a huge pile of rocks half filling the hither side: they were probably dashed from the roof in the great earth quake of 1811. Taking the left hand branch, the tourist is soon brought to "Croghan's Hall," which is nine miles from the mouth, and is the farthest point explored in that direction. The "Hall" is fifty or sixty feet in diameter, and perhaps thirty five feet high, of a semicircular form. Fronting the visitor, as he enters, are massive stalactites, ten or fifteen feet in length, attached to the rocks like sheets of ice, and of a brilliant color. The rock projects near the floor, and then recedes with a regular and graceful curve or swell, leaving a cavity of several feet in width, between it and the floor. At intervals, around this swell, stalactites of various forms are suspended, and behind the sheet of stalactites first described, are numerous stalagmites, in fanciful forms. The curious often bring some of them away. On the last occasion of our visit to the Mammoth Cave, we saw one which bore a strong resemblance to the horns of a deer, being nearly translucent. In the centre of this hall a very large stalactite hangs from the roof, and a corresponding stalagmite rises from the floor, about three feet in height and a foot in diameter, of an amber color, perfectly smooth and translucent like the other formations. On the right is a deep pit, down which the water dashes from a cascade that pours from the roof. Other avenues could be most likely found, by sounding the sides of the pit, if any one had the courage to attempt the descent. While in the part of the cave we have been describing, the visitor is far enough away from terra supra; and his dinner, which he has left at the Vineyard, claims his attention. He hastens back to the Rocky Mountains, and takes the branch which he has left on his right hand, when emerging from the Cabinet. Pursuing the uneven path for some distance, he reaches Serena's Arbor, which was lately discovered by one of the guides, named Mat. The descent to the Arbor seems so full of peril, from the position of the loose rocks around it, that many tourists will not venture it. Those, however, who have scrambled down, regard this as the very crowning object of interest, the *ne plus ultra* of their explorations. The Arbor is not more than twelve feet in diameter, and about the same in height, of a circular form; but is, of itself, floor, sides, roof, and ornaments, one perfect, seamless stalactite of a beautiful hue, and exquisite workmanship. Stalactitic matter, in the shape of blades or folds, hangs like a drapery around the sides, and reaches half way to the floor; while opposite the door, a canopy of stone projects. It is most chastely and elegantly ornamented, and has been truly designated a "fit resting place for a fairy bride." Everything about it seems fresh and new. But the invisible and unknown arranger has not quite completed this, his master piece; for the visitor can see the pure water trickling down its tiny channels, and perfecting the delicate points of some of the stalactites. Truly may the man exploring this portion of the Mammoth Cave, quote Holy Writ, and say, referring to the glittering things about him,

"That Solomon, in all his glory, was not like one of these."

But beautiful and extraordinary as are the formations in this avenue, it has been shorn of its lustre to some extent. Young ladies, whose elders and parents should have taught them better, have, acting on a criminal, though pretty general impulse, broken from the walls, and that too, in violation of the published rules, those exquisite productions of the Almighty Power, which required perhaps a hundred ages to perfect; and these have been destroyed in a minute.

This want of veneration and care for the beautiful in either Nature or Art, finds its agreeable antithesis in many European countries. We will mention the small cave of Adelburg, belonging to the Emperor of Austria, in substantiation of what we have said. The cave in question has been placed under the protecting care of Government. Ought not the same thing be done with the Mammoth Cave? Consider its mineralogical treasures alone, and the answer must be in the affirmative.

On his return from Serena's Bower, the visitor will pass, on his left, the mouth of an avenue, upwards of two miles in length, and of great height and width. A hall will be found at the termination. This avenue is generally supposed to be longer than any other in the Cave. Midway between the commencement and termination of Cleveland's Avenue is a huge rock. It is nearly circular, quite flat at the top, and three feet in height. This is appropriately called the "Dining Table." It is of great capacity, more than one hundred persons occasionally seating themselves around it. Here, by the aid of the guides (who are picked men, so far as general knowledge and obliging manners are regarded), the banquet is often spread, and the operation is no sooner concluded, than the guests fall to with a gusto the reader will be able to imagine.

The air of the Cave, with the circumstances attendant on its exploration, seem to inspire the visitors, and accordingly, many brilliant repartees and sayings are exchanged. The banquet is, indeed, two-fold; for while it can be called, in one respect, the "feast of reason and the flow of soul," with equal justice it can be designated, in a physical sense, as worthy of Ude himself.

In the midst of laughter and enjoyment, our friends, the visitors of this magnificent Mammoth Cave, have forgotten all past cause for grief, anticipate no evil for the future. Like the hero and heroine of Byron's latest poem, it may be said of them,

"The present, like a tyrant, holds them fast."

But, presently, the speech is broken. By what? Is there anything sufficiently potent to dispel so ecstatic a state of existence? Yes, and what is more, the agency is remarkably simple, being no more than a hint from the guide, that the river may possibly rise, and shut him and his companions up there. Then how sudden is the move! how quickly does the uplifted morsel drop from the mouth! and what a rattle there is of plates and dishes! Nevertheless, the hint of the guide is only an innocent ruse. Experience is the parent of judgment, and the quality in him is consummate. He sees the strong admiration of those near him, an admiration that promises to keep them where they are till long past midnight; and he knows the only mode whereby he can work out his intent, is an appeal to what is perhaps the strongest passion of the soul, fear. The deception is soon dispelled; not so, however, until various stumbles have been made by the visitors in, their hurry to reach the Echo. And when they do reach it, great is their joy at finding that it has not risen.

It must not be concluded that all objects of interest have been left behind. Far from it: the visitors, on their way to the hotel, have much to awaken profound sensations in their bosoms. The stillness, how deep is that! and how solemnly magnificent is the scene in all its details! What is that which bursts upon the silence? Are the rocks tumbling to pieces? or is the solid earth itself bursting into millions of fragments? These are the questions that are mentally put. But the continued echoes in the Cave explain the cause. A pistol has been fired, and like thunder among the hills, it reverberates through the thousand sinuosities of this cavemed earth.

Presently, our friends begin to have a thought or two concerning the time. They have seen so much since they entered the Cave, and so varied and powerful have been the sensations upcalled, that they cannot be expected to have any very strict idea of the hour. Of one thing, however, they are certain, it is, that one night at least has passed over their heads. Approaching the entrance, their "assurance is doubly assured; "for lo! there, afar, are the first gleamings of the new day. Fallible human nature! you have committed another mistake: that is not the rising sun; it is that mighty luminary upon his western throne. Solemnly, and, oh, how beautifully, does he sink away from vision; and with what thoughts, kindred of the scene around them, do the tourists seek the pleasant abode they quitted some twelve hours since!

In the course of our descriptive matter, we have not, we believe, said anything relative to the proprietorship of the Mammoth Cave. It is now, and has been for some time past, in the possession of St. George Croghan, Esq., son of the late Dr. Croghan. He is a resident of Louisville, and a gentleman of great enterprise. He has made many discoveries in the Cave. Some time since Mr. Croghan, while exploring an avenue of recent discovery, found a young child lying on the ground. From the bloom upon its cheek, and other indications of vitality, the worthy gentleman very naturally concluded that it had strayed into the Cave, and belonged to one of the farmers in the neighborhood. He felt the little stranger's cheek, and was perfectly astounded to find it cold. The child was dead. Mr. Croghan, as soon as he had come to this melancholy conviction, ordered the body to be carried to the hotel, that it might be claimed by the parents; but neither father nor mother came; and to add to the worthy gentleman's surprise, after the lapse of twenty four hours, there was nothing left but ashes. This circumstance must be taken as a striking proof relative to the purity of the air in the Cave. Anything placed there, even the body of a deceased man, woman or child, on which decay, under ordinary circumstances, is always the speediest to lay its "effacing fingers," preserves the freshness of life; but submit it to the atmosphere without, and then how swiftly is the great fiat of the Almighty fulfilled!

"Ashes to ashes, dust to dust."

We feel that we should be neglecting one part of our duties, were we to omit the enumeration of the advantages that are connected with a visit to the Mammoth Cave, Kentucky. Here we append them.

No accident more serious than an occasional stumble, causing little pain, but not so much as a discoloration, has ever been known to occur here.

Visitors, both on entering and departing from the Cave, are not likely to contract colds. They are, on the contrary, usually dispelled by a visit to this mighty and beautiful natural object.

In no part of the Cave is there to be found air of the slightest impurity.

In the Cave, a boast made erroneously by the Irish nation has been fully realized, for there no reptile has ever been found. It seems as though a charm prevailed there, to bid them away. Neither do quadrupeds trouble the Cave with their presence.

In every part of the Mammoth Cave, combustion is perfect.

Decomposition, and its loathsome adjunct, putrefaction, have never been found here.

The water of the Cave, exquisitely pure, is generally fresh; and there are, besides, one or two sulphur springs.

There are two hundred and twenty six avenues in this magnificent Cave; forty seven domes; eight cataracts; and twenty three pits.

The temperature of the place is 59° Fahrenheit, and remains precisely the same, winter and summer.

No sound, not even the loudest peal of thunder, can be heard beyond the distance of one quarter of a mile in the Cave.

We have mentioned in a former part of our details, the many testimonies that have been borne by different persons to the surpassing beauty and grandeur of the Mammoth Cave. The following are to be found in the book kept at ithe neighboring hotel:

"I have stood near the summit of Etna, watched the working waters of the Maelstroom, seen the glaciers of Iceland, visited the Giant's Causeway, stood on Termination Rocks, at the Niagara Falls; but never saw anything that could come up to the

Mammoth Cave, in extent, beauty, grandeur, and variety. J. A. G."

Query. Where are your eyes? Why, hundreds come up to it every year.

Among the prose testimonies (we shall insert some poetical ones) of which the Mammoth Cave has been the subject, we shall be pardoned the insertion of the following. The first letter, now in the hands of the publishers, is, it will be perceived, from that liberal and accomplished woman, Lady Emmeline Stuart Wortley, who was recently a visitor here, and whose impressions of the United States have just been issued from the press.

"NEW YORK, Nov. 20th, 1850.
"My DEAR SIR

"The hurry of my arrangements previous to my departure for England, will not permit me to say 'good-by' to your family and yourself a second time. I, however, trust that the day is not far distant when we shall meet again. I have already conversed with you relative to my opinions of the United States, so far as its government and social life are concerned; I have also, I believe, expressed (I fear unworthily) the admiration which its natural beauties have upcalled within my very 'heart of hearts.' Among these let me instance the Mammoth Cave in Kentucky. This is a place which all tourists should visit, so magnificent, extraordinary, and gigantic in its several details, as, in itself alone, to well repay a voyage across the broad Atlantic. During my short but pleasant stay in America, I was a frequent visitor at the Cave; and in every succeeding occasion I found fresh cause for admiration. Though a humble and trusting believer in the truths of Christianity, I am, nevertheless, persuaded that the Almighty has impressed certain characteristics on some of His works, to prove His existence and to vindicate His power, to the children of men; and never, I can safely say, were those evidences more strikingly displayed than in the Mammoth Cave.

"Let every person visit it who can; and at the same time, not forget the home attractions of the neighboring hotel. I, with many others, have tested them: no report, written or oral, can do them justice.

I am, Dear sir,

Yours very faithfully, EMMELINE STUART WORTLEY."

Some circumstances of considerable interest were connected with this amiable lady's visit to the United States. She is the daughter of the venerable Duke of Rutland, between whom and the Hon. Daniel Webster, a warm friendship has long existed. Both at the time the 'Glorious Daniel' was in Great Britain and since, the Duke of Rutland has been warmly pressed by him to pay a visit to America; but though the promise to do so has been given, it has not been realized, solely because official duties on one side and the encroachments of age on the other have stood in the way. The daughter, knowing this to be the case, felt that she could be the proxy of her parent, and gratify a wish she had long entertained, at the same time. She came to the United States, was a sharer of Mr. Webster's hospitalities, and has since returned to England to do what few British tourists of rank have hitherto done, namely, to speak favorably of the Western Republic.

The annexed letter is from another English personage of high rank and talents.

"BOSTON, June 3d, 1851.

"MY DEAR SIR:

"Last week I had the pleasure of exploring, with several other persons, the Mammoth Cave in Kentucky. I make use of the term 'pleasure' in no conventional sense, but in its true and honest one, as significant of a happy feeling.

"I have described many things in my life, with, I believe, some force and capacity, but were you to offer me the world, I could not, either to my own or your satisfaction, describe what I have seen in the monarch of caves from which I have just come.

"I can but speak in the way of likeness or analogy. Well, then, I must say that in the Mammoth Cave is a cathedral in which any Lord Primate might be proud to preach, did not the solemnity of the place admonish all humanity to humbleness, a cathedral whose gigantic buttresses and delicately wrought friezes, quatre foils, &c., are respectively as bold and minute as any ever found in the great Christian Temples of the upper world.

"There are portions of the Cave called 'Arbors,' a fit title; fanciful in their arrangement, and beautiful in their colors, they are fit for the reposing places of the prettiest fairies that ever danced in a ring beneath the moonlight. Then there are chambers of Cimmerian gloom, quite chaotic; but which, when they catch the reflection of light, display stalactites of brilliant hues, hanging as it were in mid air: and afar overhead, a second firmament studded with stars.

"We are taken along a river named after, and like what the imagination may conceive of the fabled Styx. How silent is all upon the shores of this river. How appropriate it seems, as a flood running between two worlds! Then there are Domes, large and beautiful; avenues of miles' length; chambers, waterfalls, and recesses. Verily they must all be seen to be understood.

"The neighborhood itself is a delightful one. Among its chief recommendations are the comfort and courtesies tourists from all nations can find at the Cave Hotel. He, indeed, must be a grumbler, and out of sorts with the world, who cannot make himself at home there.

"Yours ever, very truly,

"G. P. JANES."

We conclude our letters with the following. It is from an Italian painter of high repute; who, after giving to the world, by the aid of his pencil, many of the exquisite scenes of his own land, has come to America, for the express purpose of studying its features of pictorial sublimity.

"Louisville, May 14th, 1851.

"DEAR FRIEND:

"The Mammoth Cave in Kentucky is, indeed, a wonder. I have visited it day after day, and can safely assert that I never saw anything in Nature to surpass it, in features of true grandeur and sublimity.

"I repeat, that my imperfect knowledge of the language spoken in this country, will not permit me to discuss more clearly than I do, the scenes which this extraordinary Cave presents; but now I think again, I do not regret my imperfection on that point, for I feel, that were I a native, I could not speak as the reality demands I should.

"There is mutuality between all things great and beautiful: thus, a picture is often painted to the imagination by tones of music, and we hear rare melody by means of a picture. These analogies are perfect in the Mammoth Cave; and besides, great judgment seems to have been exercised in naming the different divisions. "The Bottomless Pit" and "Serena's Arbor" were, to me at least, striking examples of this. The first, indeed, seems to be deep beyond the plummet's reach; the other is fanciful and beautiful to an extreme. The Domes, too, what shall I say of them? they are full of grandeur.

"Before I close my letter, I will relate an anecdote to you, in connection with this Mammoth Cave. I had returned from my tenth visit, and was quietly seated at my supper beneath the hospitable roof of the Cave Hotel, when who should come suddenly upon me but my old friend ____. He questioned me about what I had seen, told me he should commence his own visits to the Cave next morning. So an hour passed, when, starting up, exclaimed, "But I don't know wether this Cave can be worth exploring or not, since you have so little to say about it. The truth is that I felt, as I have stated myself here to be, incapable of giving anything like a proper description of the wonders the intelligent guides had pointed out to me. I said, in reply to my friend, "Wait till tomorrow morning, when you will be able to judge for yourself." He did so, and when he came back in the evening, and also on several evenings subsequently, he was less profuse of words then I had been, though his admiration, I am convinced, was equal to my own.

"I need not speak to you of the ready attention and urbanity of the gentleman who holds sway at the Cave Hotel. You have experienced them, and I hope you will inform other persons of them.

"Yours, very truthfully,

"MARCO POGLIANI."

POETRY

The following poetical attestations of the Cave's wonders and beauties, are perhaps not the worst specimens that have been published. They are from the pens of different tourists, who, it will be perceived, have chosen their own particular subjects:

ON THE ENTRANCE TO THE MAMMOTH CAVE

To what new world are we now tending?

Is it one of pain and sin?

Whither are our footsteps wending?

How darkness hems us in!

That stream of water falling quick

How ominous its tone:

We pause, we tremble, we are sick,

And feel deserted, lone.

The water falleth in a pit,

Like tears of those who weep;

Far over head deep gloom doth sit,

And Nature is asleep.

On, on we go, soon cheer'd in soul,

For we are told that light,

Bursting from stern night's control,

Is near to trance our sight.

ON THE GRAND GALLERY

With roof magnificent, like mighty minds,

(So manifest in their commandings)

Thou standest, naught of aid from things

Lesser than thyself, demanding.

No hand of mortal man hath trac'd upon

Thy borders, forms of such rare beauty,

As to our raptured sense there now presents,

And unto which we bow in duty:

Aye, bow, and lowly, too, for GOD hath grav'd

On thee, much that of Himself doth tell;

And storm defying, they remain for e'er,

Proof even against Time's potent spell

ON AUDUBON AVENUE

Silent and cloud-like, yonder roof seems moving,

Gray in its hue, and awful in its form:

What can it be? a sign of Heaven's reproving?

Or herald of the approaching storm?

Neither; and soon we know, for high aloft,

Quick catching light, that burns nor scorches,

A multitude of gems, of colors soft,

Meet the full flame of lighted torches.

How beautiful the scene it is as tho'
The Voice omnipotent which made us all,
Had spoken once again, and bade light flow,
To banish hence for e'er, night's ebon pall.

ON THE LITTLE BAT ROOM

Where is the man can count them
Those gloomy looking things,
Lying in torpidity,
Till Spring expands their wings?
Can they e'er awake to life,
They seem so very. dead?
Can they cleave a way thro' air,
Where flow'rs and winds are wed?
Yes: of death and after-life,
Significant are they
We lie torpid in the earth,
Till the Eternal Day.
Oh, 'tis well to ponder o'er
Things like those now near us
Tho' they speak of life and death,
Still their end 's to cheer us.

ON THE CHURCH

The mind-directed hand of man
Rare wonders oft hath wrought;
To piles of beauty and of power
High treasure he hath brought.
In the Old World there now are seen
Great monuments of Art,
Gazing on which the soul expands,
And purer grows the heart.
Ancient Churches of renown,
Where the grand organ's swell,
And partner'd voices make us think
We 'mong the angels dwell;
When the stain'd glass of ages past,
Presenteth to the sight,
The Life and Death of Him we love,
In hues of Heav'n's own light.
When tombs of those who, in their day.
Walked lordly on the earth,
Are close beside the humble beds
Of lowly peasant worth.
How beautiful must service be
Performed in such a fane!
How all unlikely men should feel,

Beneath its shadows vain!

But look around, and say if aught

Of man's work hath appeared,

So truly great, and so sublime,

As this which Gon hath rear'd.

Hence we require no organ tone,

With other sounds, to give

Due sanctity to scenes around

Faith's self doth with us live.

No blazon'd scutch'on here's requir'd

No banner old and torn:

The thoughts that elevate us here,

Are not by service worn;

The song that's sung, the prayer that's breath'd,

Inspired by love's own flow,

In accents of unstudied force,

To GOD, up gushing go.

ON THE MUMMY FOUND IN THE CAVE

The tenure of our stay on earth is fleeting,

Tho' in extravagance of our conception,

We take for stable what is but a reed,

And often find, too late, that all's deception.

How petty are our aims of life! how fickle

The friends we put our trust in, and most estimate!

How glowing are the hues we paint our hopes in!

How more than earth can yield of happiness our fate!

Oh, we need monitors to tell us truly,

The one great secret of our world-creation,

That knowing all there is to know, our spirits

May not succumb to gloom or consternation.

Ponder well over that mummy form, so stark,

And it will tell thee that 'tis best, alone to fix

The inmost soul where, thron'd above the Heavens,

Sits ONE in whom eternity and love do mix.

Oh, who, beholding thee, to earth returning,

Will prate of older lineage with a brow elate:

Let thy remembrance keep him dumb, thou Mummy,

How many hundreds of his series dost thou out-date.

ON THE LOVER'S LEAP

Beautiful women, beware

Where your glances fall,

Else they may from manly hearts

Tender feelings call;

And if it should happen so,

Love met no return

That frost and snow alone were where

Summer's sun should burn;

Then how terrible might be

The mishaps of those

The disappointed Nightingales

Of a froward rose.

Let not such discarded wights

Come to Mammoth Cave,

If they have a wish to live,

Or their limbs to save.

Like the knell of Banquo brave,

Manifest to sight,

The "Lover's Leap" alas!

Will Macbeth "invite."

ON THE GREAT BALL ROOM

Shame on thy elaborations,

Artificial man;

What is all thy genius?

'Tis circled by a span.

Often, often do we hear

Of thy splendid meetings

Where exchang'd are lovely smiles,

And affection's greetings.

Satins, silks and gems are there

Music, too, cloth breathe;

While the richest flow'rs of earth,

Cunningly ye wreathe.

But, tho' all is splendid there

Luring to the eye,

And the air comes fitfully,

Like a perfumed sigh;

Here we do behold a scene

Ye never could achieve,

And in which the fancy ne'er,

Save seeing, could believe.

Fairy footsteps well might fall

On this sounding floor,

Tender vows of constancy,

Fairies here might pour.

Oh, how well can we imagine

A scene so gorgeous here!

Shunning views of lordly life,

Which the spirits sear.

ON THE GIANT'S COFFIN

"Let there be light!" such were the words

God utter'd long ago;

When darkness like a murky cloud,

Descended quick and low.

And now it seems His voice again,

The mandate high had given,

For see! how beautiful the scene,

As fair as yonder Heaven;

Where'er we gaze around, we see

Gems multiplied, and beaming

Thro' what was naught but darkness late

Like hope 'mid anguish gleaming;

Let us but trust, that when oppress'd

By pain, by care or sorrow,

We may have made manifest,

Such a glorious morrow

The likeness of the change that now

Charmeth the soul and eyesight;

Let us believe it will be so,

As sunrise followeth night!

ON THE STAR CHAMBER

How beautiful yon sky appears,

How bright each twinkling star;

And look, a comet we now see,

With awful form, afar.

'Tis now in Heaven; we see it not

Where we so late did look.

What does it mean? 'tis harbinger

Of deeds that once earth shook!

Of that dark time when hands were dy'd

In blood, and carnage sway'd;

When millions rush'd into the fight

To win or die, as bade!

Is it a star like that of him,

Who, in his mad career,

Was looked on by his fellow men

With wonder or with fear?

Is it like that

(ascendant now, Large, and outshining all),

Destin'd to a dark reverse,

Doom'd to a sudden fall?

Aye, no; no firmament is that

Ye gaze on, high, above,

No stars are they now twinkling,

Like eyes that watch for love.

'Tis the great Cave's firm roof itself,

That doth attract the gaze,

Those are its variegated gems

Of never fading blaze.

ON THE FAIRY GROTTO

Some Fairy of the olden time
Her dwelling sure had here,
And here she rested, with no grief
To shade her spirits clear.
How fit the place for one like her
So fanciful and light
A creature jocund as the dawn,
And as the morning bright.
A spot like this did he, the Bard
Of Avon's flowery stream,
Imagine, where, in "phrenzy fine,"
He had his wanton "Dream."
Titania here might move along,
And Puck his frolics play,
And Hermia in her race of love, Outpace the hours of day.
The fancy sees them passing now, How beautiful they seem!
And now they're gone, we have them not,
They've vanish'd in the gloom.

ON THE MOUNTAIN

As we go onward, the o'er-arching dome
Seems to move with us every step we take,
Like to the sky, when on the earth we roam,

113

And find few things about, the view to break:

Or like accusing Conscience, hunting one

Who, after he has done some deed of ill,

Fearing a capture, public ways doth shun,

Fancying the Avenger near him still.

Then in the profound of gloom we mingle,

And feel not of the living earth, nor near

Our fellows' habitations, single

In sense, we stand: all, all around how drear:

But light, the talisman of life and man,

Bursts cheerfully, oh, cheerfully to view:

Glory sits thron'd within the caverned span,

And all seems beautiful, and proud, and new.

Domes of the far renowned Romish pile,

And ye, Sophia and ye, too, preacher Paul,

What are your glories unto those that smile,

On this huge "Mountain," and about us fall!

ON THE RIVER SCENE

Sure we are tending to some world of doom,

And from our own have gone forever;

How very awful is this circling gloom!

How loudly sounds that one word, never!

Never! we hear it still. Alas! we know

All of our pleasant ways are ended;

No more shall music o'er our senses flow

Darkness and horror here are blended.

Or, when light glimmers o'er the wide expanse,

(Its stream, and arched rocks revealing),

Or concentrating in a point the glance,

We see red eyes upon us gleaming.

What are they, those red eyes of fearful glare,

Now shining o'er the dim, dark wave?

Do they belong to fiends, who, dwellers there,

Are foes to Him who every blessing gave?

And does the neighboring stream,

to which we come

Anon, Oblivion's waters bring?

Drinking from it, shall all of memory's sun

Be banish'd hence, like bird upon the wing?

ON SERENA'S ARBOR

Well did the monarch wise of old,

Declare if e'er himself enthron'd,

That, spite his ermine and his gold

There sprung no flow'r that would have own'd,

With aught of pride, affinity

With such a painted thing as he.

115

He said the daisy labored not,

Nor did the lily spin,

But that he knew no king who'd got

Gems of those flow'rs the twin.

No earthly diadem could be

Rarer than they in royalty.

Then, human grandeur, check thy pride,

Man lowly bow thyself,

Since simplest Nature can deride

Thy words of time and pelf.

On scenes like this, thou need'st but gaze

To purify thy heart and ways.

THE END.